RFAR
W9-AOT-169

ASTRONOMY AND CULTURE

Greenwood Guides to the Universe
Timothy F. Slater and Lauren V. Jones, Series Editors

Astronomy and Culture
Edith W. Hetherington and Norriss S. Hetherington

The Sun
David Alexander

Inner Planets
Jennifer A. Grier and Andrew S. Rivkin

Outer Planets
Glenn F. Chaple

Asteroids, Comets, and Dwarf Planets
Andrew S. Rivkin

Stars and Galaxies
Lauren V. Jones

Cosmology and the Evolution of the Universe
Martin Ratcliffe

ASTRONOMY AND CULTURE

Edith W. Hetherington
and Norriss S. Hetherington

Greenwood Guides to the Universe
Lauren V. Jones and Timothy F. Slater, Series Editors

GREENWOOD PRESS
An Imprint of ABC-CLIO, LLC

A B C CLIO

Santa Barbara, California • Denver, Colorado • Oxford, England

Library of Congress Cataloging-in-Publication Data

Hetherington, Edith W.

Astronomy and culture / Edith W. Hetherington and Norriss S. Hetherington.

p. cm. — (Greenwood guides to the universe)

Includes bibliographical references and index.

ISBN 978-0-313-34536-4 (hard copy : alk. paper) — ISBN 978-0-313-34537-1 (ebook)

1. Ethnoastronomy. 2. Archaeoastronomy. 3. Stars—Mythology. 4. Astronomy—Folklore. I. Hetherington, Norriss S., 1942– II. Title.

GN476.3.H48 2009

520.89—dc22 2009007368

13 12 11 10 9 1 2 3 4 5

This book is also available on the World Wide Web as an eBook. Visit www.abc-clio.com for details.

ABC-CLIO, Inc.
130 Cremona Drive, P.O. Box 1911
Santa Barbara, California 93116–1911

This book is printed on acid-free paper ♾

Manufactured in the United States of America

For
Bob White
January 28, 1947–December 1, 2007
brother and brother-in-law
who taught our son Robert how to play poker

Contents

Series Foreword

Not since the 1960s and the Apollo space program has the subject of astronomy so readily captured our interest and imagination. In just the past few decades, a constellation of space telescopes, including the Hubble Space Telescope, has peered deep into the farthest reaches of the universe and discovered supermassive black holes residing in the centers of galaxies. Giant telescopes on Earth's highest mountaintops have spied planet-like objects larger than Pluto lurking at the very edges of our solar system and have carefully measured the expansion rate of our universe. Meteorites with bacteria-like fossil structures have spurred repeated missions to Mars with the ultimate goal of sending humans to the red planet. Astronomers have recently discovered hundreds more planets beyond our solar system. Such discoveries give us a reason for capturing what we now understand about the cosmos in these volumes, even as we prepare to peer deeper into the universe's secrets.

As a discipline, astronomy covers a range of topics, stretching from the central core of our own planet outward past the Sun and nearby stars to the most distant galaxies of our universe. As such, this set of volumes systematically covers all the major structures and unifying themes of our evolving universe. Each volume consists of a narrative discussion highlighting the most important ideas about major celestial objects and how astronomers have come to understand their nature and evolution. In addition to describing astronomers' most current investigations, many volumes include perspectives on the historical and premodern understandings that have motivated us to pursue deeper knowledge.

The ideas presented in these assembled volumes have been meticulously researched and carefully written by experts to provide readers with the most scientifically accurate information that is currently available. There are some astronomical phenomena that we just do not understand very well, and the authors have tried to distinguish between which theories have wide consensus and which are still as yet unconfirmed. Because astronomy is a rapidly advancing science, it is almost certain that some of the concepts presented in these pages will become obsolete as advances in technology yield previously unknown information. Astronomers share and value a world-view in which our knowledge is subject to change as the scientific enterprise makes new and better observations of our universe. Our understanding of the cosmos evolves over time, just as the universe evolves, and what we learn tomorrow depends on the insightful efforts of dedicated scientists from yesterday and today. We hope that these volumes reflect the deep respect we have for the scholars who have worked, are working, and will work diligently in the public service to uncover the secrets of the universe.

Lauren V. Jones, Ph.D.
Timothy F. Slater, Ph.D.
University of Wyoming
Series Editors

Preface

Conscious contemplation of the thoroughness with which astronomy is embedded in the cultures of great civilizations is found as far back in history as the Roman Empire. Cicero (106–43 BC) wrote that the necessary and utilitarian arts, like agriculture and house building, were followed in parades of civilizations by the arts that civilized people pursue in their leisure time, including astronomy, poetry, rhetoric, and philosophy. These bring human beings together in society and join them by the bounds of speech and writing.

Astronomical allusions and references are woven throughout Western civilization's cultural heritage. A single, illustrative example may suffice here. In *Romeo and Juliet*, Shakespeare references the changing phases of the Moon:

> O swear not by the Moon, th'inconstant Moon
> That monthly changes in her circled orb
> Lest that thy love prove likewise variable.

The phases of the Moon, both their continuous change and their constant repetition, are a widely recognized common sight, not withstanding the fact that many people now spend far more time under roofs than under the sky. Even poets cannot miss the scientific pith.

Other facets of astronomy are more arcane. Why Saturn, Jupiter, and Mars are nearer to the Earth at opposition than at conjunction very likely is better understood by astronomers than by cultural historians specializing in Renaissance thought.

Historians, however, more likely will appreciate how astronomical facts resonated with philosophical currents of Copernicus's time and how the appearance of an admirable symmetry and clear bond of harmony in the universe could have constituted a compelling argument for the Copernican system. Especially when rival theories could not be separated by observations, human values guided scientists' choices.

Interrelationships between astronomy and culture are found throughout history and in endless abundance. Yet the sciences and the humanities purportedly have developed contradictory attitudes and values.

The medical diagnosis is schizophrenia. Indeed, this psychotic disorder supposedly has progressed so far that humanistically illiterate scientists (so they are crudely stereotyped) and scientifically illiterate humanists (to maintain both the iteration and the vituperation) reportedly are openly antagonistic and unable to communicate with one another.

Should any such schizophrenics actually exist, tragically scientists and humanists each would be unable to know, connect to, and interact with the other, complementary half of their whole, integrated culture. Schizophrenics would be cut off from the totality of their creativity and intellect.

That might not be so bad, were culture merely the alien and exotic behavior of other people to be perused, probed, and pondered by scholars of our superior civilization. But culture also pertains to us. Culture's many components include aesthetics, behaviors, beliefs, biases, conventions, customs, desires, emotions, fads, fashions, forms, habits, inclinations, manners, methods, modes, mores, paradigms, patterns, practices, predispositions, preferences, styles, superstitions, tastes, traits, values, and ways.

Culture is more than a little arbitrary; it is even irrational at times. Astronomy, in contrast, generally is thought of as a strictly rational enterprise. Any humanistic tendencies toward imaginative speculation are increasingly constrained by an underlying factual universe increasingly discoverable by observation and reason.

Thus, at least at first glance, the chasm between astronomy and culture may seem unbridgeable. However, a closer look finds that astronomy and culture are inextricably intertwined in our civilization. In the pages of this book, we have attempted to illustrate a few instances of the intricate interrelationships of astronomy and culture that make both the richer. If we have failed, it is our failure to seek out and document, not a failure of our society to include astronomy within its general culture.

We wish that our college astronomy courses, way back when we were students, had at least hinted at a few of the many fascinating interrelationships between astronomy and culture. Furthermore, one of us wishes that he had included more cultural content in the astronomy and history of astronomy courses he taught. It is our fervent hope that the following pages may open to readers the wide and enjoyable world of astronomy and culture.

The enormous diversity of connections between astronomy and culture is hinted at in the collection of illustrations obtained for this book by Kevin Downing of ABC-CLIO. Their discovery and acquisition presented far more of a challenge than normally encountered in the production of a typical book. With a modest budget, but a wealth of energy and ingenuity, Kevin made this book possible. We also wish to acknowledge the magnificent History of Science Collections at the University of Oklahoma Libraries. One of us (he, again) has had the privilege of teaching in the History of Science Department there and working in the collections. In the rare instances that an illustration in a published book was not already scanned and available from Oklahoma, Professor Kerry Magruder supplied it for this book.

1

Archaeoastronomy

Prehistory is the time before written records, continuing in some civilizations long after others have begun committing to paper or its equivalent their commercial and civic dealings, their religions, and their philosophies. Archaeologists attempt to learn about prehistoric peoples and their cultures from whatever remains of a past civilization, from something as small as a piece of broken pottery or as large as the ruined buildings of an extended empire. In one branch of archaeology, archaeoastronomers search prehistoric ruins for astronomical alignments, speculate on their possible purposes (often for religious ceremonies), and attempt to extrapolate from the level of astronomical sophistication to overall cultural development and achievement. Their conclusions are far from certain, but archaeoastronomy nevertheless has provided exciting new suggestions about ancient civilizations.

STONEHENGE

Now hailed as the eighth wonder of the ancient world and the source of much speculation about cultural achievements of early Britons, Stonehenge also illustrates the tortuous path of archaeoastronomical discovery and understanding.

Julius Caesar (100–44 BC) twice invaded Britain, but not until AD 43 did Rome finally prevail. Marching inland, the conquerors discovered gigantic stones, or megaliths (Greek: *mega* = large and *lithos* = stone). Reaching twenty feet upward and weighing a hundred thousand pounds each, they

Figure 1.1: Stonehenge. Drawing from Charles Knight, *Old England: A Pictorial Museum of Regal, Ecclesiastical, Baronial, Municipal, and Popular Antiquities* (London: C. Knight & Co., 1845).

were capped by other gigantic stones, some hanging over the edges of the vertical stones. Thirty upright megaliths formed a circle over a hundred feet in diameter, with a continuous lintel around the top spanning the openings. The Romans were little impressed by Stonehenge (hanging stones, from Old English: *hengen* = hung up or hanging)—nor were they impressed by the native inhabitants.

Legends grew up around Stonehenge. Merlin, King Arthur's magician, supposedly used his magic to disassemble the Giant's Ring in Ireland (moved there from Africa by giants), ship it to Britain, and reassemble it to commemorate slain British noblemen. Or, perhaps a Celtic warrior queen had erected Stonehenge as a monument to herself. Or, maybe Stonehenge was a Druid temple. Not until 1740 did anyone notice that the main axis of Stonehenge points to midsummer sunrise, and another century and a half passed before the British astronomer Sir Joseph Norman Lockyer (1836–1920) probed Stonehenge's astronomical alignment.

Astronomical Alignments

Vacationing in Greece in 1890, Lockyer noticed alterations in temple foundations, presumably to realign them. He also knew that east windows of some English churches face sunrise on the day of their patron saint. Might

Greek temples have been similarly aligned with astronomical phenomena, Lockyer wondered. He borrowed a pocket compass and began taking measurements. Data on Egyptian temples was also available to Lockyer, and he found at Karnak, 400 miles upstream from modern Cairo and established around 2000 BC, temples oriented toward sunset at summer solstice, the longest day of the year, and temples near the Great Pyramid of Giza oriented toward sunrise and sunset at the spring and fall equinoxes, when day and night are of equal length. Lockyer unhesitatingly concluded that ancient Egyptians worshipped the Sun, although solar alignments of English churches had not led him to the same conclusion regarding his own countrymen.

The Sun's Motion

The Earth's axis of rotation is not perpendicular to the Earth's orbit around the Sun, but is inclined at an angle of about 23.5 degrees. The Sun's apparent annual motion carries it alternately above and below the plane of the Earth's equator. The solstices are the times of the year when the Sun is at its highest or lowest latitude as seen from the Earth. Summer solstice, or midsummer, occurs about June 22; winter solstice, or midwinter, occurs about December 22. In the United States, these days are regarded as the beginning of summer and winter. Astronomically, however, they mark the middle of summer and the middle of winter, and are so understood in Europe. In the Southern Hemisphere, the times of summer and winter are reversed.

The equinoxes are the points on the celestial sphere where the plane of the Earth's orbit around the Sun and the plane of the Earth's equator intersect. Vernal or spring equinox occurs about March 21, and autumnal or fall equinox about September 22. At these times, the length of day and night are equal (twelve hours) everywhere on Earth.

In his 1895 novel *The Purchase of the North Pole*, the French science fiction writer Jules Verne (1828–1905) concocted a fictional scheme to get at coal beneath the Arctic ice. The mine would be brought into the temperate zone by changing the inclination of the Earth's axis with a tremendous explosion.

Scientists now believe that something similar may actually have happened to the Moon, where a giant impact crater near the equator may have rolled to the south pole. (A spinning bowling ball similarly stabilizes with its finger holes, the area of least mass, at the axis of rotation.) The tilt of the Earth's axis of rotation may vary over time, and with it global cooling and precipitation patterns, possibly causing cycles of extinction of plant and animal life.

Aligning an Altar, Poetically Recounted

The 1822 poem *To the Lady Fleming, On Seeing the Foundation Preparing for the Erection of Rydal Chapel, Westmoreland* by William Wordsworth (1770–1850) has masons waiting until sunrise on the day of the patron saint to align an altar:

Then, to her Patron Saint a previous rite
Resounded with deep swell and solemn close,

Through unremitting vigils of the night,
Till from his couch the wished-for Sun uprose.
He rose, and straight—as by divine command,
They, who had waited for that sign to trace
Their work's foundation, gave with careful hand
To the high altar its determined place;

Puritans attempted to eradicate the practice of honoring patron saints, and the chapel of Emmanuel College, Cambridge was aligned north-south.

Other Christian groups also aligned their churches astronomically. Saint Peter's Basilica in Rome is aligned so on the vernal equinox the rays of the rising Sun pass through opened outer and inner doors and fall on the high altar.

New temples constructed on much older foundations, but differently oriented, suggested to Lockyer stellar alignments, which also change slowly over time, with the stars moving around the sky in a 26,000 year cycle. Lockyer studied Egyptian temples built about 2525, 1250, and 900 BC, and found that their alignments differed by the amount necessary to keep up with stellar precession.

In another temple, an inscription stated that on the first day of the new year, the star Sirius shone into the temple, mixing her light with that of her father, the Sun, on the horizon. Lockyer calculated that the temple had pointed toward Sirius around 700 BC. The same date was also calculated from positions of stars on a medallion found at the temple.

Back in England, Lockyer calculated astronomically that Stonehenge had faced sunrise at summer solstice around 1680 BC, a date roughly confirmed from flint hammers and axes dated to about 1800 BC. Archaeologists, however, were not impressed; they preferred that astronomers stick to astronomy and leave the archaeology to professional archaeologists.

Stonehenge Decoded

Not until the 1960s would an astronomer persuade skeptics that Stonehenge was an astronomical observatory. Gerald Hawkins (1928–2003) grew up in England only vaguely interested in Stonehenge, until he went to work at a nearby missile-testing base and was able to visit Stonehenge frequently. After moving to America and beginning a book on astronomy, his thoughts focused on Stonehenge.

Perhaps, Hawkins speculated, some of the stones, archways, potholes, and mounds at Stonehenge marked astronomical alignments. He chose 120 pairs of points and determined to which part of the sky each pair pointed.

With new computer technology, he was able to calculate in a minute a mass of probable alignments that would have taken months to do by hand. Not surprisingly, as many as a dozen lines were associated with the Sun. To his surprise, Hawkins also found a dozen lunar alignments. The number of correlations seemed too great to be coincidental.

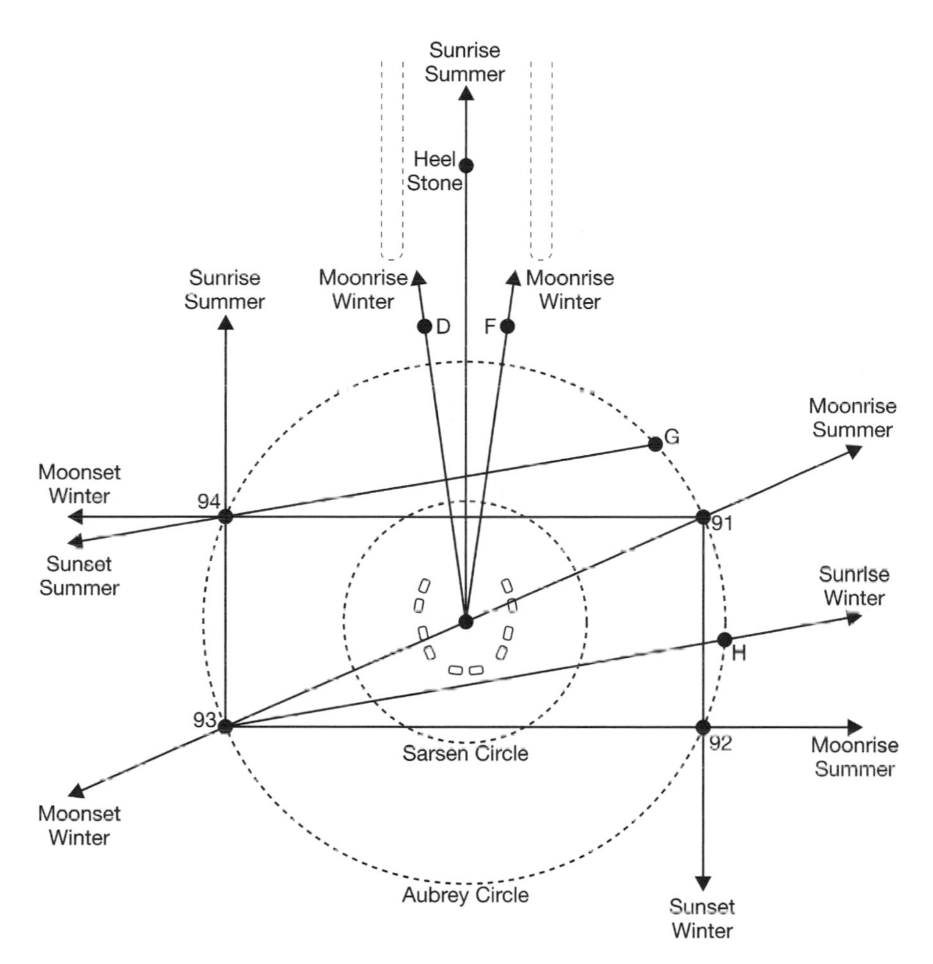

Figure 1.2: Schematic Plan of Stonehenge. Seen from the center of the Sarsen Circle along the main axis and avenue, the Sun dramatically rises over the heel stone at midsummer. A line drawn through stones 93 and 94 also points toward midsummer sunrise, while a line through stones 93 and H points toward midwinter sunrise. Midsummer and midwinter sunsets are marked by lines through stones G and 94 and through 91 and 92. Lines through 93 and 91 and through 93 and 92 point toward midsummer moonrise. Lines from the center toward D and F mark midwinter moonrises, while lines through 91 and 93 and through 91 and 94 point toward midwinter moonsets. From Gerald S. Hawkins in collaboration with John B. White, *Stonehenge Decoded* (Garden City, NY: Doubleday, 1965). Illustration by Jeff Dixon.

Stones of Stonehenge

The largest stones, forming the Sarsen Circle and weighing up to a hundred thousand pounds each, are sandstone from the Marlborough downs, 18 miles to the north. Stonehenge's igneous bluestones, tens of thousands of pounds each, are similar to rocks in Wales, 240 miles away. In 2004, workmen laying a pipe near Stonehenge uncovered bodies whose teeth, under chemical analysis, revealed that the men had grown up in Wales; perhaps this little band of workers had helped bring stones to Stonehenge. A year later, in 2005, archaeologists found a site in Wales with many worked stones lying around; perhaps this is where Stonehenge's bluestones were quarried. An alternative theory is that glaciers carried masses of rock from Wales to Stonehenge.

Ancient Astronauts?

Hawkins also applied his computer methods to the Nazca Lines in Peru, where more-recently exposed gravel, even if uncovered a thousand years ago, is lighter in color than areas longer subject to the Sun's rays. There is little rainfall—for only twenty minutes a year—and little wind on the high desert to erase markings. Hawkins examined 186 lines, but found no more suggestion of astronomical alignments than pure chance would have produced. So much for the theory that the lines are a giant astronomical calendar predicting the seasons and propitious times for planting and harvesting, or that the lines are a giant map left by ancient astronauts, maybe even marking a landing strip for their return. Perhaps people walked along the lines in prayer to their gods for water, or intended the lines as gifts to their gods.

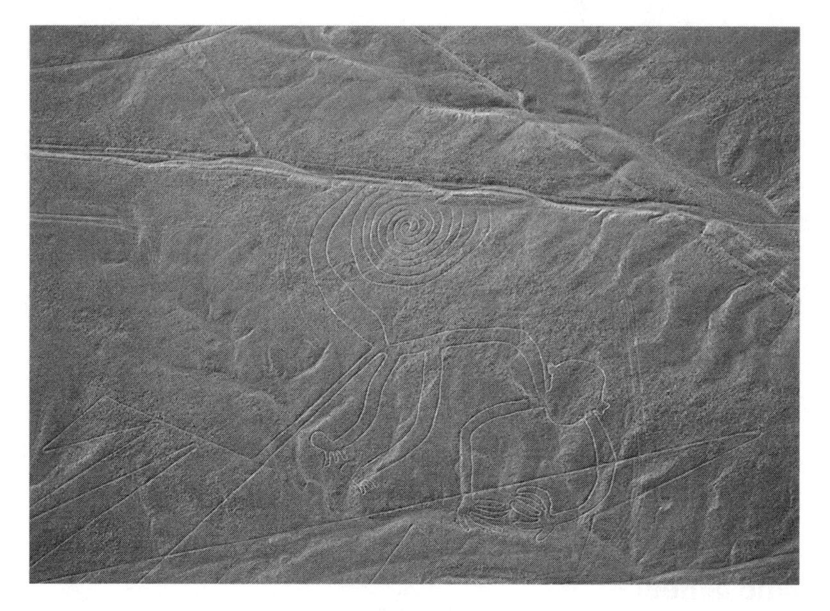

Figure 1.3: Nazca Lines and Figures. Satellite photograph of straight lines and figures scratched into the Nazca Desert of Peru. The lines stand out in contrast against undisturbed surface darkened by long exposure to sunlight. The figures are called geoglyphs (Greek: *ge* = Earth and *glyphein* = to carve). Near upper-center of this photograph is a spiral, and below it a monkey. Archaeoastronomers have searched for astronomical alignments, so far without success, and the purpose and significance of the lines and figures remains a mystery. Shutterstock image 8243188; Copyright Jarno Gonzalez Zarraonandia.

Thoughts about Stonehenge

Perhaps ancient Britons, lacking writing, set stones to record their observations of solar and lunar positions on the horizon at different times of the year. They probably recognized periodicities and predicted when celestial phenomena would reappear; Stonehenge was their giant calendar. Stonehenge could also have been an attempt by megalomaniacs to immortalize themselves; Egyptian pharaohs built pyramids, and in our own time builders reach for the sky and immortality with skyscrapers.

The recognition at Stonehenge of the cyclical nature of astronomical phenomena, remarkable as it was, does not prove that ancient astronomers of Stonehenge had any general scientific theory about motions and causes, from which astronomical data could have been deduced mathematically. Stonehenge does testify, however, to the existence of a stable society and culture capable and motivated to observe the heavens over many generations and sufficiently organized to build the massive monument.

Difficulties inherent in procuring, setting up, and rearranging gigantic stones make it unlikely that the final version of Stonehenge seen today was the working observatory where solar and lunar positions were measured and recorded. Such tasks would have been done more easily during an earlier phase of Stonehenge, in a more easily manipulated and modified Woodhenge or Timberhenge. Stonehenge is a clock commemorating the passage of time and a museum commemorating a great scientific achievement. It may also be a sacred site, where people conducted religious rituals connected with astronomical phenomena.

ASTRONOMICAL ALIGNMENTS IN THE NEW WORLD

Once the first astronomical alignments were demonstrated, enthusiastic and vigorous searches for more began. Inevitably, the quest spread from the old world to the new.

The Big Horn Medicine Wheel

In 1974, an American solar physicist turned archaeoastronomer in his spare time examined a Native American medicine wheel, so named for its location on a ridge of Medicine Mountain in Wyoming's Big Horn Range. The wheel consists of rocks arranged on the ground in a circle seventy-five feet in diameter with twenty-eight rows of rocks (the same number used in roofs of ceremonial lodges) radiating out from the center to the circumference. Several piles of rocks (cairns) are located just outside the wheel. Similar wheels are found elsewhere in Wyoming, South Dakota, Montana, and Alberta and Saskatchewan, Canada. Radiocarbon dating of a bone from

Figure 1.4: Big Horn Medicine Wheel. A line from the cairn (pile of rocks) in the front right of the photograph along the adjoining spoke and through the center of the wheel would point to midsummer sunrise on the far horizon. From the next cairn to the left, lines can be drawn through the center of the wheel pointing to heliacal risings of the stars Aldebaran, Rigel, and Sirius. Photograph by Richard Collier, Wyoming Department of State Parks and Cultural Resources.

under the central rocks of one wheel yields a date of around 2500 BC. The Big Horn wheel is thought to date from between AD 1200 and 1700.

The Big Horn wheel may have an alignment with midsummer sunrise. More doubtful, but not impossible, are alignments with the heliacal risings of the stars Aldebaran, Rigel, and Sirius. (The Greek god Helios drove the Sun's chariot across the sky, and *heliacal* means relating to or near the Sun. The *heliacal rising* of a star, a planet, or the Moon is the first day that the object is visible for a brief moment on the Eastern horizon at dawn, before it is lost sight of in the Sun's glare.)

Sirius rising at the end of August would have warned of the end of summer and that it was time to leave the mountain. The wheels also might have guided nomadic Indians on hunting trips. And there is the usual suspect: religious ceremonies might have been conducted at the sites.

Towers of Chankillo

The oldest solar observatory yet found in the Americas was recognized as such only in 2007, although archaeologists had known about the site for decades. Occupied as early as the fourth century BC, Chankillo on the coast of Peru contains within its massive walls many plazas and buildings, including a fortified temple. Running north-south for a thousand feet along the ridge of a nearby hill are thirteen towers, believed to be about 2,300 years

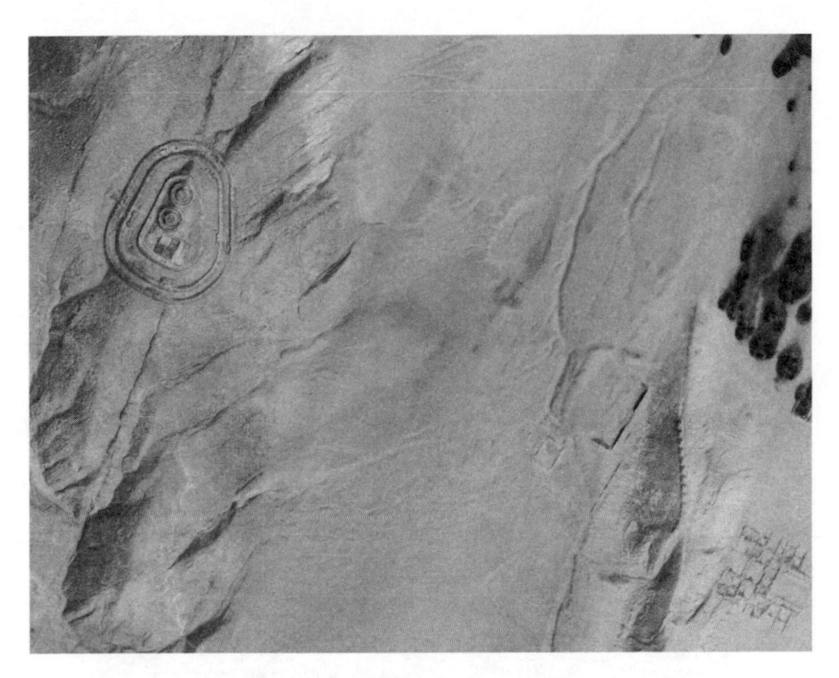

Figure 1.5: Chankillo Observatory. Satellite photograph of the 2,300-year-old "Norelco ruin" in Peru, so-named for the resemblance of its central complex (upper-left) to a modern electric shaver. On another ridge, thirteen towers (slightly below and to the right of the center of the photograph) are aligned with solar events. Near the towers is a storage complex (lower right). Satellite image courtesy of GeoEye/SIME. Copyright 2008. All rights reserved.

old. At summer solstice (December in Peru), the Sun rises to the right of the rightmost tower, and at winter solstice, the Sun rises to the left of the leftmost tower.

Written records dating back to the Spanish conquest in the sixteenth century state that the Incas constructed markers for positions of the Sun at important times of the year. Sun pillars, now long vanished, are said to have been built on the horizon near Cuzco, supposedly to indicate the right time to plant crops. Spaniards also documented state-regulated Sun worship by the Incas, with the Inca king claiming to be an offspring of the Sun. Perhaps the Thirteen Towers at Chankillo enabled kings to foretell motions of the Sun and thus convince their people that they controlled the Sun.

Chaco Canyon's Sun Dagger

In 1977, Anna Sofaer, an artist studying petroglyphs (Greek: *petros* = rock; *glyph* = carving) in New Mexico's Chaco Canyon, noticed that near noon a beam of sunlight passing between two large rock slabs bisected a spiral-shaped symbol carved on the cliff; she also noticed that the day was summer solstice. Returning at fall equinox, Sofaer observed a smaller sun dagger slicing through another smaller petroglyph nearby, while a large sun dagger cut

Figure 1.6: Chaco Canyon Sun Dagger. Photograph at noon on summer solstice of a beam of sunlight bisecting a spiral petroglyph after having passed between two large rock slabs. Copyright: Solstice Project. Photo: Karl Kernberger

through the large spiral halfway between its center and its outer edge. Furthermore, at winter solstice, two parallel daggers bracketed the large spiral.

Before the Chaco Canyon petroglyphs were discovered, archaeoastronomy was limited to searches for astronomical orientations; now it includes patterns of light and shadow on rock paintings and carvings. Once one example was discovered, more were soon found at other locations.

The sun dagger helps confirm cultural contacts between ancient cultures, suggested by the presence of trade items including sea shells and macaw feathers at Chaco Canyon. The larger spiral at Chaco Canyon tracks the Moon's eighteen-year cycle, but no evidence has yet been found that other American Indians knew about the lunar cycle; the Maya did know.

After the 1982 film *The Sun Dagger* was televised, the National Park Service closed the area to all but scientific researchers. Erosion at the site had already begun, however, and one of the large rock slabs funneling sunlight onto the petroglyphs has since pivoted out of position. Chaco Culture National Historical Park, home to the largest pre-Columbian ruins yet discovered in the United States, has been named a world heritage center by the United Nations Educational, Scientific, and Cultural Organization (UNESCO), in recognition of its great cultural value and the urgent need for its protection.

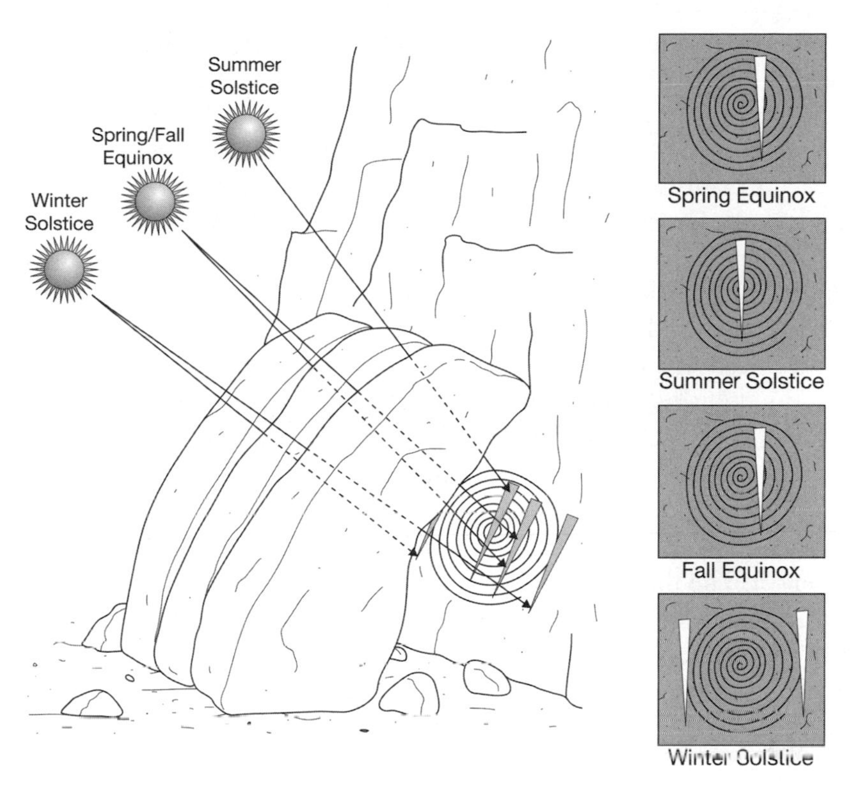

Figure 1.7: Seasonal Sun Daggers. Diagram showing how beams of sunlight passing between large rock slabs bisect or bracket in different ways at spring equinox, at summer solstice, at fall equinox, and at winter solstice, a large spiral petroglyph incised on the canyon wall behind the stone slabs. After Thomas Y. Canby, "The Anasazi: Riddles in the Ruins," *National Geographic* (November 1982). Illustration by Jeff Dixon.

CONCLUSION

If one were plopped down in the middle of an ancient civilization by a malfunctioning time machine, the best way to establish communication with the natives might well be to arrange rocks pointing to sunrise at the midsummer solstice. A common curiosity about appearances on the horizon of the Sun, Moon, planets, and stars has existed over an amazingly broad range of historical eras, geographical locations, and cultures.

Literary invention, too, testifies to a fascination with appearances of celestial objects. In *The Hobbit*, J. R. R. Tolkien (1892–1973), professor of Anglo-Saxon and English literature at Oxford University and author of the *Lord of the Rings*, imagined that the little people of his fictional Middle-Earth shared the human interest in astronomical alignments. Early in the story, a map is introduced containing the advice to stand by the gray stones when the thrush knocks on Durin's Day, because then the setting Sun will shine upon the keyhole. As all Middle-Earthers knew, Durin's Day was the first day of the last moon of autumn on the threshold of winter, when the

Moon and the Sun were in the sky together. Unfortunately, it was beyond Middle-Earthers' ability to calculate in advance the date of the next occurrence of this auspicious day. Amid songs of farewell, the hobbit Bilbo Baggins and his companions ride off the next morning, a midsummer morning as fresh and fair as could be dreamed. Later, with autumn far along, Bilbo's group reaches Lonely Mountain, towering grim and tall before them. There they find the secret door mentioned in the map. But no matter how much they beat on it, thrust and push at it, and implore it to move, nothing stirs. The next day, the beginning of the last day of autumn, sees Bilbo gazing gloomily westward through a narrow opening framed by rocks. The orange ball of the Sun sinks toward the rim of the Earth, near which also appears a thin new Moon, pale and faint. A thrush lands near Bilbo. Meanwhile, the Sun becomes obscured by clouds. Suddenly, when hope is lowest, a red ray from the Sun comes like a finger through an opening in the clouds, and through an opening in the rocks, and falls on the smooth stone face of the secret door. The thrush trills. With a loud crack, a sliver of rock flakes away, uncovering a hole in the secret door. "The key! The key that went with the map!" cries Bilbo. "Try it now!"

Writers of fiction are free to imagine in great detail the cultures of their invented worlds, after which astronomy can be called into service to advance the story. Archaeoastronomers begin with the astronomy, and from it attempt to infer something about the underlying culture, hopefully more truth than fiction. The search for astronomical alignments among ruins of ancient civilizations is now a well-established and generally accepted scholarly activity. Also generally accepted is the belief that astronomical accomplishment implies additional intellectual achievements, although specific correlations between astronomical and other cultural accomplishments are yet to be established. The future will see additions to knowledge of astronomical activities among ancient cultures, and new and improved thinking about ancient cultures.

Playing Tricks on the Future

The lobby wall of an office building in Vancouver, British Columbia, Canada, is graced by a glass-and-steel sculpture twenty-seven feet wide, twelve feet high, and four feet deep. A label reveals that *Navigation Device: Origin Unknown*, purportedly found fragmented and eroded in 1990 on Lyell Island, British Columbia, is a fictitious device. Should this work of art become separated from its label in the future, will archaeoastronomers then ponder its purpose and the civilization that created it?

RECOMMENDED READING

Aveni, Anthony. *Stairways to the Stars: Skywatching in Three Great Ancient Cultures* (New York: John Wiley & Sons, Inc., 1997).

Hawkins, Gerald S. in collaboration with John B. White. *Stonehenge Decoded* (Garden City, N.Y.: Doubleday, 1965).

Lockyer, Norman. *The Dawn of Astronomy: A Study of the Temple-worship and Mythology of the Ancient Egyptians* (London: Cassell, 1894).

———. *Stonehenge and Other British Monuments Astronomically Considered* (London: Macmillan, 1906; expanded 1909).

North, John. *Stonehenge: A New Interpretation of Prehistoric Man and the Cosmos* (New York: Free Press, 1996).

Ruggles, Clive. *Ancient Astronomy: An Encyclopedia of Cosmologies and Myth* (Santa Barbara, CA: ABC-CLIO, 2005).

Sofaer, Anna. *Chaco Astronomy: An Ancient American Cosmology* (Santa Fe, NM: Ocean Tree Books, 2008).

FILMS

Cracking the Stone Age Code. 1971. 50 minutes. Color. BBC-TV. Produced by Paul Johnstone.

Secrets of Lost Empires – Stonehenge. 1997. 60 minutes. Color. Nova television documentary series. Produced by Cynthia Page and Julia Cort. Written by Julia Cort.

Shadows of the Ancients (Anasazi astronomical alignments and markings). 2001. 29 minutes. Color. KUAT television series *The Desert Speaks*, episode 1309. KUAT Communications Group, University of Arizona. Host David Yetman.

The Sun Dagger. 1982. 58 minutes. Color. Produced by The Solstice Project. Directed by Anna Sofaer and Albert Ihde. Narrated by Robert Redford.

The Mystery of Chaco Canyon. 1999. 56 minutes. Color. Produced by The Solstice Project. Directed by Anna Sofaer. Written by Anna Sofaer and Matt Dibble. Music by Michael Stearns. Narrated by Robert Redford.

WEB SITES

International Center for Archaeoastronomy: http://www.wam.umd.edu/~tlaloc/archastro/index.html.

NASA. *Ancient Observatories: Timeless Knowledge*: http://sunearthday.nasa.gov/2005/locations/chaco.htm.

The Solstice Project: http://www.solsticeproject.org.

2

Mythology

Human fascination with celestial phenomena, manifest in astronomical alignments of prehistoric ruins, is also a foundation of astronomical myths. These stories about bright objects in the sky moving regularly in patterns discernible over time reflect people's lives, render the world and humankind's place in it intelligible, and legitimize, rationalize, and validate social institutions and cultural traditions. Celestial allusions and references are woven into literature, art, and music, and people familiar with them delight in enhanced comprehension, appreciation, and enjoyment of what they read, view, and listen to.

People try to understand what they observe in the sky. The Sun, especially, rising, moving across the sky, providing light and heat, and then setting, commands a prominent place in the myths of ancient cultures. Absent our modern understanding that a body in motion remains in motion until acted upon by an outside force, ancient peoples either required a supernatural force (such as a god) to move the Sun, or imagined the Sun as a living being (again, a god) moving under its own power. The Moon, too, was the subject of myths, its waxing and waning dramatically visible and calling for explanation.

The same astronomical phenomena are observed in different cultures, and, as is only to be expected, myths from different civilizations share some similarities. However, explanatory expectations and the meanings attributed to astronomical phenomena also are reflections of different cultural values, and consequently, differ across cultures. Myths cannot be completely construed as inevitable responses to observational facts, but must also be analyzed and understood as cultural constructs.

EGYPT AND MESOPOTAMIA

Egypt enjoyed a nearly continuous civilization and culture for more than three thousand years, from the unification of Upper and Lower Egypt in 3110 BC to the Roman conquest and the death of Cleopatra in 30 BC. Mythology evolved along with political and cultural unification: the gods of several villages were combined into hybrids, and gods of more important cities became more important gods. When the princes of Thebes became rulers of all Egypt, they unified their local god Amon with Ra, the sun god, and made Amon-Ra the king of the gods. Sobek, the crocodile god, already a coalescence of the elemental gods of earth, water, air, and fire, became Sobek-Ra and was depicted with a sun disk over his head.

In one of the first known instances of monotheism, the Egyptian pharaoh Akhenaten declared in 1344 BC that Aten, formerly a variant of the sun god Ra, now was the only god. The American composer Phillip Glass (b. 1937) celebrated Akhenaten's religious beliefs in his 1983 opera *Akhnaten*, as he had the science of Albert Einstein (1879–1955) and the politics of Mohandas Gandhi (1869–1948) in earlier operas about people with inner visions who altered their cultures.

Figure 2.1: Aten, the Sun. Limestone relief, fourteenth century BC, Amarna, Egypt. The Egyptian pharaoh Akhenaten (also known as Amenophis IV, reigned 1353–1336 BC) is depicted with his chief wife Nefertiti (ca. 1370–1330 BC, and perhaps pharaoh after Akhenaten and before Tutankhamun) and two children worshipping the Aten, the visible Sun. Photograph by Jürgen Liepe.

Some Egyptian rulers sought political stature through a cult of personality by associating themselves with the Sun. Queen Hatshepsut (d. 1458 BC) claimed that Amon was her father; and Rameses II (ca. 1303–1213 BC, and probably the pharaoh from whom Moses demanded release for his people), claimed to combine the powers of his father, of Amon, of Ra, and also of Set, god of the desert, and of Horus, whose two eyes were sometimes called the Sun and the Moon.

The Eye on the Dollar Bill

The Egyptian god Horus lost an eye in a fight, and Thoth, the god of magic and of the Moon, restored Horus's eye: the eye was dark and then restored, an allusion to the phases of the Moon. Horus's all-seeing eye became a sign of protection in Egypt, was adopted as a symbol by Freemasons and Rosicrucians, and is featured on the reverse side of the Great Seal of the United States of America, atop a pyramid with thirteen steps, one for each of the original thirteen colonies, and with the date 1776 (in Roman numerals), the year of independence, on the bottom step. In 1935, the seal was placed on the back of the one-dollar bill.

Other rulers, as well, have attempted to link their images with that of the Sun. In Rome, the Emperor Caracalla (186–217) portrayed himself as the son of the sun god. In Peru, the conquering Incas brought out their ruler at a distance and wearing a gold cloak reflecting sunlight, to convince less sophisticated conquered tribes that the Inca, their ruler, was the child of the Sun. In a period of increasing French power and influence, Louis XIV (1638–1715) was deservedly known as Le Roi Soleil, the Sun King. In Japan, the land of the rising Sun, so displayed on the national flag, the emperor claimed direct descent from the sun goddess. Saparmurat Niyazov (1940–2006), the egomanical dictator of Turkmenistan, had a gold statue made of himself and placed on a gigantic marble apparatus, arms raised, rotating in step with the Sun and seemingly guiding the Sun across the sky. And a 2008 Chinese video documentary presented a portrait of Chairman Mao (1893–1976) with sunbeams radiating from his head.

In more a literary than a historical context, William Shakespeare (1564–1616) in his play *Richard III*, employed the Sun as a metaphorical pun to denote a son in the York dynasty:

> Now is the winter of our discontent
> Made glorious summer by this sun of York.

Transportation by boat on the Nile was the main mode of movement in ancient Egypt, and this fact is mirrored in myth: a boat moved the Sun across the sky during the day, and a second boat carried the Sun through the waters under the Earth back to the east for the next day. The Mesopotamian sun god was transported sometimes by boat and sometimes by chariot, crossing the sky during the day and making his way back through an underworld cavern at night. The Greeks, lacking navigable rivers, put their sun god in a chariot. The ten suns of Chinese mythology, which took turns lighting the sky, were pulled in a chariot by six dragons or were carried by a crow, depending upon the particular version of the myth. The ancient Hindu sun god, with hair of flames, also crossed the sky in a chariot. Lacking wheeled conveyances of their own, New World civilizations also had none for the Sun. In Navajo mythology, the sun god carried the Sun across the sky on his back and hung the Sun on a peg in his house during the night.

Sun God in a Boat

Shamash, the Mesopotamian sun god, fiery rays emanating from his shoulders and wearing a tall conical hat with four rows of bull horns indicating high status, steers a boat made of bundles of reeds waterproofed with bituminous tar. The boat's cargo includes a human-headed lion, a large jar, and a plow. The figure on the prow at the front of the boat, a lesser god with only a single pair of horns, propels the boat with a punting pole (much as Oxford University students now do on the River Thames, known where it passes through Oxford as the River Isis, after an Egyptian goddess). Behind, on shore, is Shamash's sister, Ishtar, the goddess of sexual love, fertility, and vegetation, holding plants and with grain sprouting from her body. Fish swim in the river. A dragon is coiled around the back of the boat, blending with the pattern of the reeds at the bottom of the boat. The war of the gods against the forces of chaos, represented by dragons, was a frequent subject of Mesopotamian mythology.

Figure 2.2: Sun God in a Boat. Line drawing of a printout from a Mesopotamian cylinder seal, Akkadian period, 2224–2154 BC. Bibliotèque Nationale, Paris. Illustration by Jeff Dixon.

Babylonians were interested in the purpose of celestial objects, particularly with reference to maintaining order. Marduk, the patron god of the city of Babylon and the most powerful of all Mesopotamian gods after his city became the political capital of Mesopotamia, slew the dragon Tiamat, the enemy of order. The victory was celebrated in an annual new year's festival, usually at the first new moon after the vernal equinox in spring, when crops began growing, but sometimes at winter solstice, the shortest day of the year, after which increasingly longer days symbolized rebirth.

Measurement of time is an important aspect of order, and Marduk caused the Moon to shine and appointed him to signify the days, marking off every month by means of his crown or horns (the phases of the Moon). Luminous horns signified the first six days of the month (beginning with the first visible crescent), reaching a half-crown (first quarter crescent moon) on the seventh day; on the fifteenth day both halves were visible (full moon); then the Moon diminished its crown and reduced its light; on approaching the course of the Sun, the Moon disappeared on the thirtieth day.

Other cultures less singularly focused on order also sought explanations for the lunar phases. The Maya imagined two moon goddesses, one a beautiful young woman, for the new crescent moon, and the other an old crone,

Babylonian Boundary Stone

Kudurrus, now called boundary stones (Akkadian: *kudurru* = boundary), but historically stored in temples, confirmed land contracts between king and vassal, stated a divine curse to be placed on anyone breaking the contract, and presented symbolic images of the gods protecting the contract. On the top row of this boundary stone, from left to right, are symbols for the gods Ishtar (eight-pointed star; divine personification of the planet Venus), Sin (crescent moon; moon god), and Shamash (disk with rays; sun god). In the next row are two trees, a turtle, and twisted horns; and in the third row are a draghead, a sphinx, and a scorpion. Some speculations attribute to these symbols astronomical as well as theological significance, possibly recording a comet traveling through constellations Draco the sky dragon, Leo the lion, looking like a sphinx, and Scorpio, but the Babylonian zodiacal constellations were not established until after this kudurru was carved. Speculation also has it that the snake, which runs up the left side of the slab and ends at the top in the head of a bird, represents the Milky Way and the star Cygnus.

Figure 2.3: Babylonian Boundary Stone. Stone slab sculpted with images in low relief, nine inches long and five inches wide, circa 1100 BC. © The Trustees of the British Museum.

for the full Moon. Shakespeare used the changing phases of the Moon as a contrast to the constant Sun in *Henry V:*

> But a good heart, Kate, is the sun and the moon; or rather the sun, and not the moon; for it shines bright and never changes, but keeps his course truly.

Babylonian concern with order is also evident in their myths concerning the zodiac, a band of stars centered on the ecliptic, the path of the Sun across the heavens as seen from the Earth. Marduk determined the year, divided it into twelve months, and set up three constellations (arbitrary configurations of stars, often imagined to mark outlines of animals) for each month. The paths of the gods Anu, Enlil, and Ea determined the bounds of the stars in the zodiac, so none would err or go astray.

With the subordination of gods under Marduk, Babylonia's polytheistic mythology and religion took a major step toward monotheism, and likely influenced subsequent Judeo-Christian beliefs. Beginning around 1000 BC in Jerusalem, Hebrews subordinated all other gods to Yahweh. Next, their god lost his human form and foibles, and evolved into a more abstract vision. In Judeo-Christian beliefs, as in Babylonian mythology, while the Sun is given a place in the creation story and is set as a light in the heavens, there is little attempt to explain the Sun's observed daily movement across the sky. It is the creator, not the Sun, who is important in Marduk myths and in Genesis. When Joshua commands the Sun to stand still to ensure the Hebrews' victory against their enemies, the faithful are meant to be impressed with the power of the creator, not left wondering how it was done.

THE NEW WORLD

In addition to the motions of the Sun and the Moon and any order they might mark, maintenance of these celestial objects was also a concern in some cultures. In Chinese mythology, the Sun and the Moon received a weekly dusting to make them bright again. Considerably more was required by Aztecs and Maya to keep the Sun shining. An animating spirit existing in human blood and hearts supplied energy to the Aztec sun god, and supposedly some 80,000 prisoners were sacrificed at the Great Pyramid of Tenochtitlan (now the location of Mexico City) in 1487 (although that works out to an implausible fourteen per minute for four days and nights continuously, with a single stone knife). Four priests held down the subject on an altar on a raised temple, while a fifth priest with an obsidian knife made an incision below the ribs and pulled out the still-beating heart, which was then burned. Frequent wars of conquest provided a steady stream of sacrificial victims. The Maya also nourished the gods with their own blood, piercing their tongues, ears, extremities, and genitals.

Coyolxauhqui, Aztec Moon Goddess

Coyolxauhqui, the Aztec moon goddess, was also associated with blood and killing. When her mother became pregnant yet again, Coyolxauhqui encouraged her four hundred sisters and brothers to kill their mother. Huitzilopochtli, a god of war, emerged from his mother as a fully armed adult and sprang to her rescue. He cut off Coyolxauhqui's head and threw it into the sky, forming the Moon. Every month thereafter she died, was cut into pieces, and reborn in phases.

An Aztec monument depicts Coyolxauhqui decapitated and her arms and legs dismembered and oozing blood. Her head (at the top, facing upward) is adorned with a feather headdress, her hair with circles, and her cheek with bells. Around her waist is a two-headed serpent belt, holding a skull on her back (her breasts are shown frontally, but her hips are turned and seen in profile). The serpent is repeated on her arms and legs. The sandals on her feet have a mask with fangs.

Figure 2.4: Coyolxauhqui, Aztec Moon Goddess. Circular volcanic stone slab, approximately ten-and-a-half feet in diameter and eight tons in weight, found by electricity company workers installing underground cables in Mexico City in 1978. Museo del Templo Mayor, Centro Histórico, Ciudad de México.

Some Mayan monuments show their moon goddess seated on a crescent moon holding a rabbit, whose profile the Maya observed on the face of the Moon. In myths, a rabbit was thrown into the face of Tecuciztecatl (who became the Moon), dimming his brightness relative to the Sun's. There are also Chinese and Japanese myths about a rabbit in the Moon, and just before the 1969 Apollo 11 Moon landing, Houston asked the crew to look for a large Chinese rabbit.

In contrast to centrally controlled Egypt and also to the Incan and Aztec empires in Mesoamerica, the Maya were more a loose confederation of city-states, and this political organization was reflected in a variety of solar and lunar deities in different locations and at different times. After the Spanish conquest, the Mayan moon goddess was fused with the Virgin Mary in some local practices.

Possibly even more important for the Maya than either the Sun or the Moon was the planet Venus. The Venus god was Quetzalcoatl and his twin, Xolotl, the evening and morning stellar appearances of the planet Venus. Their unity was a metaphor for the force that moved the Sun. As recorded in a Mayan table, Venus disappeared from view (when it was between the Sun and the Earth) for about 8 days. This period corresponds to Quetzalcoatl's eight-day stay in the underworld. Next, Venus appeared in the eastern sky as a morning star and quickly attained its greatest brightness (at its maximum angular separation from the Sun). Quetzalcoatl, resurrected, had ascended the throne as god. Venus remained visible in the morning sky for 236 days, before falling below the horizon. It remained invisible for 90 days (on the opposite side of the Sun from the Earth). This was a particularly frightening time for the Maya, who feared that Venus might not reappear. Human sacrifices were made at the first new appearance of Venus, when the planet was at its faintest (farthest from the Earth) and presumably most in need of sacrificial sustenance; this was also the time for going to war. Venus was now an evening star (on the opposite side of the Sun from when it was a morning star), and it remained visible in the night sky for 250 days. The mean period for this cycle was 584 days. Contrary to the data in this particular Mayan table, the average periods of visibility and invisibility actually are 8, 263, 50, and 263 days, with the same mean period, of 584 days. Five Venus cycles very nearly equaled eight solar years, so appearances of Venus were easily coordinated with the seasons.

In Incan mythology, the Sun took pity on people who did not have agriculture or weaving, and sent his two children, the Inca and his sister, to teach these skills. Another tale, however, reports the first myth to be just that, a myth, created by conquering Incas to convince those they conquered that the Inca was the child of the Sun. In any event, at the winter solstice, Incas gathered to honor the sun god and plead for his return. Celebrants fasted for days before the event, refrained from sexual intercourse, and presented gifts to the chief Inca. Llamas were sacrificed to ensure good crops, and the Sun was symbolically tied to a stone hitching post, called an *intihuatana*. The Incas' Catholic conquerors banned the ceremony and destroyed all *intihuatana* they found, but at least one survived, in Macchu Picchu, the hidden fortress that the Spaniards never found.

Modern Peruvians, in an effort to renew native traditions, have reinstituted the celebration, to the extent that they can imagine it from study of architectural ruins and early written accounts, though some of the more gruesome sacrificial elements are omitted. The Cuzco festival is one of the largest and most colorful in South America; reportedly, more than a hundred thousand people, both natives and visitors, gather for music, dancing, and pageantry.

The Inuit, indigenous to Northeast Canada, believe that when the world began, Anningan lived with his sister Malina in a large village. One night, while Malina was dancing and singing, a wind blew out all the lamps, and in the dark a man pushed Malina down and raped her. Fearing that the man might return, she blackened the palms of her hands with soot. When the wind again blew out the lamps, the man came again, and this time Malina marked him with her sooty hands. When the lamps were relit, she discovered that the man was Anningan. Malina took a torch from the wall and ran out into the night. Taking up another torch, Anningan followed her, but he tripped, and his torch was reduced to a few embers. A windstorm lifted them both high into the sky, where Malina, pregnant with Anningan's child, gave birth to the Earth. Malina, bright with her burning torch, was turned into the Sun; Anningan, pale, with only a few embers, became the Moon. Anningan continues to chase Malina across the sky. Forgetting to eat, he grows thinner; when he becomes hungry enough, he disappears for three days each month, the moonless nights of the new moon, to eat. To avoid her brother, Malina tries to stay opposite him, on the other side of the Earth; thus the Sun and Moon rise and set at different times. Occasionally, Anningan catches Malina, and again rapes her; thus are eclipses explained. Given the prominence of the Moon during an uninterrupted darkness lasting for weeks north of the Arctic Circle, it is not surprising that it was prominent in Inuit culture and myth.

The mythology of the Ge people in Brazil also features a Sun-Moon rivalry, though it is more comedy than tragedy. When they were roasting meat, the Sun grabbed all the fat pieces, the Moon complained, and the Sun responded by throwing a hot piece at the Moon, burning him and causing the spots now visible on the Moon.

GREECE AND ROME

Functions of myths include characterizing the universe in familiar terminology and categories, explaining natural phenomena, offering moral lessons, and providing entertainment. The best myths do all four. Greek and Roman myths, particularly, also are the source of many astronomical allusions and references permeating Western literature, art, and music.

The Sun

Each morning Helios, the sun god, drove the Sun's chariot up the arc of heaven and across the sky to descend in the west at night. When he rashly swore an unbreakable oath to give his son Phaethon anything the boy asked for, Phaethon demanded to drive the chariot. (A light, four-wheeled horse-drawn vehicle is now called a phaeton.) Helios explained the dangers involved, but the proud young man refused to listen to his father. The youth, unable to control the supernatural horses, allowed the celestial chariot to swing low, endangering the Earth, and Zeus shot Phaethon down with a thunderbolt—as depicted by the Flemish painter Peter Paul Rubens (1577–1640), the French painter Gustave Moreau (1826–1898), and others.

This myth describes the Sun in everyday language, explains a natural phenomenon, offers a moral lesson (that rash oaths lead to disasters and rash young men who do not listen to their fathers come to bad ends), and is entertaining.

There is also a Chinese myth involving the Sun and headstrong youth whose threat to the Earth is ended by their death. Ten suns took turns traveling across the sky, but on one occasion they all entered the sky together, scorched the Earth, and killed people. Nine of them were shot down; the tenth saved himself by hiding, and came out only when called by the rooster, as the Sun still does.

Myths, suitably altered, are still employed to inculcate moral lessons. Chinese youth are taught that justice will overcome evil. Originally, Kua Fu chased after the Sun to see where it went at night, and he died from thirst. Somewhat similarly in Greek mythology, Icarus rashly flew too near the Sun and his wings, constructed of feathers and wax, melted. The Icarus legend metamorphosed into a story of aspiration and heroic audacity more suitable to support the renewed optimism of the European Renaissance, and also to commemorate with a statue of Icarus a University of Virginia student shot down in World War I. Analogously, Kua Fu's curiosity has been transformed into a mission to save the Earth from eternal darkness by catching up with the crow that carried the Sun, informing the bird that Kua Fu had defeated the wicked spirit of the mountain, and thus persuading the crow to overcome its despair and continue to carry the Sun around the sky. Kua Fu was so tired that he died, but he continues to help people, having

Sun Words

Helios, the Greek sun god's name, also forms English words. The *heliosphere* is a region or bubble around the solar system containing ionized atoms (the solar wind) blown off the Sun's outer atmosphere; the *heliospheric current sheet* is a ripple in the heliosphere caused by the Sun's magnetic field; the *heliosheath* is the outer region of the heliosphere, the solar system's final frontier, where the solar wind begins to mix with interstellar gas and dust; and the *heliopause* is the outer boundary of the heliosphere, beyond which is interstellar space. The adjective *heliacal* means relating to or near the Sun, and the *heliacal rising* of a star, planet, or the Moon occurs on the day it is first seen rising above the horizon at dawn, immediately before being lost sight of in the glare of the Sun. The Copernican *heliocentric world view* put the Sun in the center. *Aphelion* is the point in an orbit around the Sun farthest from the Sun (*apo* = away from), and *perihelion* is the point nearest the Sun (*peri* = near). A *heliometer* is a telescope designed to measure the apparent diameter of the Sun, a *photoheliograph* is an instrument for photographing the Sun, and a *heliostat* is an instrument that forms a stationary image of the Sun at the focus of a telescope. A *heliograph* is a signaling device reflecting sunlight, and the message it sends is a *heliogram*. *Heliotropism* is the orienting stimulus of sunlight turning (*tropos* = turn) *heliotropic plants*, *heliotropes*, toward the Sun. *Heliotrope* is also a color. A *helianthus* is a sunflower. *Helioaerotherapy* treats disease by exposure to sunlight. *Heliolatry* is worship of the Sun. In 1868, British astronomer Norman Lockyer, on observing a previously unknown line in the Sun's spectrum not attributable to any known element, suspected that he had discovered a new element existing only in the Sun, and he named it *helium*.

Apollo

In late Greek mythology, the god Apollo became associated with Helios and the Sun. In present-day thinking, Apollo is the apotheosis and quintessence of Greek gods, the one most often associated with the glory of classical Greece. Encountered on a distant planet by the intrepid adventurers of the television series *Star Trek*, Apollo turned out to be a member of an alien race that had visited Earth during the time of classical Greek civilization and taught culture to the Greeks in return for their worship. The National Aeronautics and Space Administration (NASA) named its first human spaceflight program Mercury, a messenger in the sky, and its second manned program, featuring a two-person spacecraft, Gemini, after the constellation with the twin stars Castor and Pollux. For the third program, Abe Silverstein, director of NASA's Office of Space Flight Programs, remembered from his grade school days the story of the god who rode the chariot of the Sun, and he chose the name Apollo, but not before going back to his old textbook and making sure that Apollo had not done anything that "wouldn't be appropriate." Silverstein didn't realize that textbooks are sanitized, and that the lustful and licentious behavior of Greek gods was methodically expurgated. A lunar goddess would have been more appropriate, given the program's goal of landing a man on the Moon, but NASA was no more ready for a feminine name for a space program than it was for a female astronaut at that time. Sally Ride did not go into space until 1983.

been transformed into a forest. Greek heroes often were transformed into constellations in the sky.

Mythological explanations often lagged behind evolving scientific understanding, although one Greek myth did have the Sun during the night providing light to people on the other side of the Earth. Uneducated Greeks may have believed in a flat Earth, but Aristotle (384–322 BC) argued that the Earth is a sphere. Ancient Greek philosophers also realized that the Moon reflected sunlight, however much in myths it shone with its own light.

The Moon

Selene is the best-known Greek lunar deity. Like her brother Helios, she drove a chariot across the sky, pulled by two horses rather than Helios's four, as befit a smaller light and a less powerful god. Selene had many lovers, usually other gods, but also the shepherd Endymion, with whom she had fifty daughters, representing the months between the Olympic games. Selene was worshipped on the days of new and full moons, and was important enough to the ancient Greeks to have inspired a poem describing her beauty and power. After encountering Greek culture, and in the process of absorbing and being absorbed by Greek culture, the Romans amalgamated names and functions of their gods with Greek mythology, and Selene became Luna, the Roman goddess of the Moon.

In Europe, the Moon came to be blamed for inducing madness, particularly at full moon—hence the words *lunacy* and *lunatic*. Shakespeare's

Hymn to Selene

Poems created and memorized during an earlier oral tradition may have been dictated to scribes in ancient Greece between the eighth and sixth centuries BC, when the Greek alphabet was introduced. Much has been attributed to Homer, especially the epic poems the *Iliad* and the *Odyssey*, although it is far from certain that such a man even existed. The *Homeric Hymns*, a collection of joyous declarations of thanks to individual gods, once were attributed to Homer himself, and at least are Homeric in employing the same rhythmic pattern and dialect used in the *Iliad* and the *Odyssey*. Of Selene we read:

Ye Muses, sing of the fair-faced, wide-winged Moon;
ye sweet-voiced daughter of Zeus son of Cronos,
accomplished in song!
The heavenly gleam from her immortal head circles the Earth,
and all beauty arises under her glowing light,
and the lampless air beams from her gold crown,
and the rays dwell lingering when she has bathed her fair body in the ocean stream,
and clad her in shining raiment,
divine Selene,
yoking her strong-necked glittering steeds.
Then forward with speed she drives her deep-maned horses
in the evening of the mid-month when her mighty orb is full;
then her beams are brightest in the sky as she waxes,
a token and a signal to mortal men.
With her once was Cronion wedded in love,
and she conceived, and brought forth Pandia the maiden,
pre-eminent in beauty among the immortal Gods.
Hail, Queen,
white-armed Goddess,
divine Selene,
gentle of heart and fair of tress.
Beginning from thee shall I sing the renown of heroes half divine
whose deeds to minstrels chant from their charmed lips;
these ministers of the Muses.

From Andrew Lang, *The Homeric Hymns; a New Prose Translation and Essays Literary and Mythological* (London: G. Allen, 1910).

Othello, after killing his beloved Desdemona, attributed his mad action to the Moon:

It is the very error of the Moon;
She comes more nearer Earth than she was wont,
And makes men mad.

Othello's defense might have succeeded in an English court of law. In the seventeenth century, the Lord Chief Justice of England recorded that the best medical authorities of his day believed that the Moon had a great influence in all diseases of the brain, and persons were usually in the height of their distemper around the full moon, especially a full moon at an equinox or summer solstice. However, by the nineteenth century, this medical opinion was recognized as outdated. Thus a lunar lunacy defense would not have prevailed for Wozzeck, the protagonist in the 1925 opera of the same name by the Austrian composer Alban Berg (1885–1935). Wozzeck goes mad and kills his wife, just after she remarks how red is the rising Moon. Like bloodstained steel, Wozzeck replies, drawing his knife. (At a total lunar eclipse, the Moon takes on a blood-red color; the eclipse of AD 734 supposedly foretold the death of the Venerable Bede, an English monk, who did die the following year.) Wozzeck tossed his bloodstained knife into a pond, in which he also washed blood off himself, and light from the Moon turned the water blood red, incriminating him.

Mercury

The Romans, as had the Babylonians and the Greeks, associated gods with planets, and the Romans conferred on the planets the names they still command today. Mercury, the swiftest moving of the planets, was named for the Roman god of commerce and travel. Analogous to the Greek god Hermes, Mercury also was the god of writing and a swift messenger from the gods to humans. Also, he was credited with establishing months and recognizing the courses of the constellations. *Mercurius,* the Latin version of Mercury, may have been derived from *mercari* (to trade), from *merces* (wages), or from *merx* or *mercator* (a merchant).

In Mayan mythology, the underworld lords had four messengers for communicating with the upper world. All were owls, perhaps one pair corresponding to the waxing and waning of Mercury as a morning star and the other pair for Mercury's evening appearances. Owls, like Mercury, appear only at twilight, before sunrise and after sunset.

In Shakespeare's *The Merry Wives of Windsor,* Falstaff urges Mistress Quickly, his good she-Mercury, to be brief in imparting her message. And in *Richard III,* a tardy cripple bearing a countermand fails to overtake winged Mercury carrying the order to kill the king's brother. In *King John,* a messenger is commanded:

> Be Mercury, set feathers to thy heels,
> And fly like thought from them to me again.

And in *Troilus and Cressida*:

> A Grecian and his sword, if he do set
> The very wings of reason to his heels
> And fly like chidden Mercury from Jove.

Messages are carried by the Astronomical Society of the Pacific's *Mercury*, a popular astronomy magazine, and by the San Jose *Mercury-News* newspaper. Speed is the implied message of the Bristol Mercury aircraft engine and the Mercury automobile. Speedy messages are the product of: the Mercury Mail Transport System, an e-mail server; the Mercury Messenger, an instant messaging program; and Mercury Communications, a telecommunications company. Flowers are speedily delivered by a floral delivery service whose logo is Mercury in winged sandals and cap carrying a bunch of flowers.

Postal workers have adopted as their symbol Mercury's winged staff with two entwined snakes. The caduceus is sometimes confused with the legendary ancient Greek doctor Asclepius's wingless staff with a single serpent entwined around it. *Star Trek*'s Starfleet medical headquarters had a stylized version of Mercury's staff, and the U.S. Army Medical Corps and the U.S. Naval Hospital Corps also mistakenly use Mercury's symbol.

The Columbia Business School uses the symbol, too, referencing Hermes-Mercury as the protector of merchants. Both the Mercury dime, minted in the United States between 1916 and 1945, and carved column capitals at the New Orleans Customhouse depict Mercury's head with wings on his cap.

A mercurial person, such as the hotheaded Mercutio in Shakespeare's *Romeo and Juliet*, is erratic, volatile, and unstable, reflecting Mercury's swift movements. The British composer Gustav Holst (1874–1934) in his 1916 orchestral suite *The Planets* portrayed Mercury with the ethereal sound of the Impressionists, everything in dabs and dashes of sound. Mercury has also been associated with the mind. The English poet Geoffrey Chaucer (ca. 1343–1400) wrote in *The Canterbury Tales*:

> For women are the children of Venus,
> And scholars those of Mercury; the two
> Are at cross purposes in all they do;
> Mercury loves wisdom, knowledge, science,
> And Venus, revelry and extravagance.
> Because of their contrary disposition
> . . .
> That's why no woman ever has been praised
> By any scholar.

Mercury, the metal, is liquid at room temperature, and is sometimes called quicksilver. Shakespeare's Hamlet speaks of poison coursing through the body as quickly as quicksilver. It is also the name for manufacturers of ultra light aircraft and of surf wear, for a boat built by Mercury Marine, and for a Marvel Comics superhero with the superhuman ability to run at the speed of sound (named Max Mercury in DC Comics, to avoid trademark confusion). Quicksilver Messenger Service was a popular psychedelic band in San Francisco in the 1960s.

Venus

Venus, the brightest of the planets, was named for the Roman goddess of love and beauty (Babylonian: *Ishtar*; Greek: *Aphrodite*, meaning foam-born; Chinese: *Tai-pe*, meaning beautiful white one). Venus is never far from the Sun in the sky, rising shortly before sunrise as a morning star or after sunset as an evening star. In Babylonian mythology, the planet Venus was the sister of Shamash, the sun god. In Mayan mythology, Venus and the Sun were twins.

In Roman mythology, the goddess Venus arose from the foam of the sea, and so she is pictured, wafted on a large sea shell to shore as a gift from Heaven, in *The Birth of Venus* by the Italian Renaissance painter Alessandro Botticelli (1444–1510). He was responding to a renewed interest among educated Italians in things Roman, including classical mythology. In a fresco painted on a house in Pompeii, the Roman city destroyed by the volcanic eruption of Mount Vesuvius in AD 79, Venus is depicted lying in a seashell, a metaphor in classical antiquity for a woman's vulva. Botticelli could not have seen this painting, because excavations at Pompeii only began in the eighteenth century. Nor is it likely that he saw a Roman mosaic in Tunisia also with Venus and a seashell. But he certainly knew the mythology. In other frescos at Pompeii, Venus inclines in the arms of Mars and is caressed by Mars, with Cupid present in both backgrounds. The city was full of erotic art: advertisements for services offered or intended to increase the pleasure of customers. Botticelli's painting, irreverently known as "Venus on the half-shell," inspired scenes in the 1962 James Bond film *Dr. No* and in a Monty Python skit. The Russian-American writer Vladimir Nabokov (1899–1977) compared Botticelli's Venus and the young Lolita in his 1955 novel *Lolita*.

Holst portrayed the planet Venus with a nostalgic glance back to the music of the German composer Richard Wagner (1813–1883), who told the story of legendary lovers in his opera *Tristan und Isolde*. Shakespeare mentioned Venus in *A Midsummer Night's Dream*:

> When his love he doth espy,
> Let her shine as gloriously
> As the Venus of the sky.

A myth involving Ishtar, the Babylonians' divine personification of the planet Venus, explained the observed fact that the planet disappears from the sky for long periods of time, and also explained the changing of the seasons. Ishtar descended into the underworld seeking the return of her imprisoned lover, the god of spring. At each of seven gates she had to part with an item of clothing, and she arrived completely naked. She was attacked by the plague demon, smitten with disease from head to foot, and kept prisoner. In her absence, all fertility above was lost. When her brother Shamash, the sun god, sent an order for her release, Ishtar was sprinkled with the waters of life, received back a garment at each of the seven gates, returned to the surface fully clothed, and fertility was restored on Earth.

In 1891, the Irish playwright and novelist Oscar Wilde (1854–1900) echoed the Ishtar myth in his play *Salome*, with its dance of the seven veils. Wilde also had the Moon as red as blood before the dramatic decapitation of John the Baptist. The biblical story of Salomé has her dancing for a king on his birthday, but no seven veils, nor nakedness. Public portrayal of biblical characters was prohibited in Britain, but Wilde's play was a great success in Paris and Berlin. The German composer Richard Strauss (1864–1949) made Wilde's *Salome* into an opera, and movies eventually followed, including a 1953 version starring Rita Hayworth. The English illustrator Aubrey Beardsley (1872–1898) drew images for a publication of Wilde's play.

Mars

Mars, the red planet, was named for the Roman god of war; the month of March also owes its name to this god. Yet again, Shakespeare could not avoid mentioning planets. In *Henry VI, Part I*:

> Mars his true moving, even as in the heavens
> So in the Earth, to this day is not known:
> Late did he shine upon the English side;
> Now we are victors; upon us he smiles.

Holst portrayed Mars, the bringer of war and an aggressor, with a low, menacing melody, which later accompanied the ominous Death Star in the 1977 movie *Star Wars*.

Men are from Mars and women are from Venus, so the saying goes, the former characterized by aggression and the latter by romance and love. In "The Wife of Bath's Tale" in his *Canterbury Tales*, Chaucer combined planetary influences in the woman of the tale, her disposition given by both Venus and Mars at her birth:

> Certainly I am all Venerian
> In feeling; and my heart is Martian.
> Venus gave me my lust and lecherousness;
> And Mars gave me my sturdy hardiness.
> Taurus was my birth-sign, and with Mars therein.
> Alas, alas, that ever love was sin!
> And so I always followed my inclination,
> By virtue of my constellation
> That made me that I could not withdraw
> My chamber of Venus from a good fellow.

Jupiter

The planet Jupiter was named for the king of the gods (also known as *Jove*; Greek: *Zeus*; Babylonian: *Marduk*). Jupiter was the patron of the Roman state, and in Shakespeare's *Antony and Cleopatra*, Julius Caesar is called the Jupiter of men. Jupiter was also in charge of cosmic justice, and ancient Romans swore to him in their courts; hence the expletives "by Jupiter!" and "by Jove!" Zeus-Jupiter ruled Mount Olympus, residence and sporting place of the gods, when they weren't fooling around elsewhere on Earth, and Jupiter was known as a boisterous and happy god, involved in numerous escapades and dubious love affairs with goddesses and mortals both. Hence the words *jovial*, *joyous*, *jolly*, and *jollity*. Holst characterized Jupiter musically with a jolly English folk song.

Saturn

The planet Saturn was named for the Roman god of agriculture, and the day Saturday also owes its name to this god. Over a whole week in December, the Saturnalia festival licentiously celebrated Rome's agricultural prosperity with food, drink, and merriment.

Yet *saturnine* means cold and steady in mood, slow to act or change, and of a gloomy or surly disposition. That's because the Greek god Chronos, ruler of time and often depicted as Father Time, was defeated by the Olympians, left Greece for Italy, changed his name to Saturn, became a king, and taught his new people how to farm. Association of Chronos-Saturn with despondency, despair, melancholy, miserableness, and wretchedness follows from notions of old age, limitations, and death. Holst represented the planet with a musical procession winding relentlessly to its end. Appropriately, Saturn is the slowest of the planets known before the invention of the telescope.

Marsilio Ficino (1433–1499), a humanist philosopher of the Italian Renaissance, in his 1489 *De vita libri tres* (Three Books on Life) wrote that outstanding individuals were under Saturn's astrological influence and inclined to melancholy. The German artist Albrecht Dürer (1471–1528) in his etching *Melancolia* transplanted this concept of melancholy from mythology and astrology to the visual arts, and for the next several centuries artists throughout Europe produced variations on Dürer's theme. Shakespeare remarked on Saturn's sway over human character in *Much Ado About Nothing*. Don John, who cannot hide what he is, wonders that Conrade can act benevolently, contrary to Saturn's influence:

> I wonder that thou, being, as thou sayest thou art, born under Saturn,
> goest about to apply a moral medicine to a mortifying mischief.
> I cannot hide what I am.

And in *Titus Andronicus*:

> Madam, though Venus govern your desires,
> Saturn is dominator over mine:
> What signifies my deadly-standing eye,
> My silence and my cloudy melancholy.

The metal lead—dull, leaden, and heavy—was associated with the planet Saturn by alchemists. In medicine, *saturnism*, named for its symptoms, which include headaches, fatigue, irritability, and depression, or melancholy, reminiscent of a saturnine humor, is now known to be the result of lead poisoning.

Constellations

People dream of reaching for the stars, but many fewer stars, perhaps as few as one percent of those seen in ancient times, now are commonly visible. Light pollution is the culprit, from streetlights, buildings, and billboards in densely populated areas, a side effect of industrialization. The International Dark-Sky Association, started by amateur and professional astronomers, is trying to reclaim the night sky.

The Big Dipper, as Americans perceive it, is one of the most recognized star patterns in the sky. It is an asterism, a small arbitrary grouping of stars; larger and also arbitrary groupings of stars are called constellations. Early Britons imagined the Big Dipper as a cart, and it is still envisioned as a wagon by Germans and Scandinavians. Britons now see it as a plough, and occasionally, especially in Northern England, as a butcher's cleaver. In southern France, it is discerned as a saucepan. The Starry Plough banner, with the seven stars of the Big Dipper, is a political symbol in Ireland, and the state flag of Alaska features the Big Dipper plus Polaris, the North Star, to which the two stars at the end of the Big Dipper upwardly point. Africans knew the Big Dipper as a drinking gourd, and runaway slaves in the United States followed it north to freedom. Observers with extraordinarily good eyesight can discern that the middle star in the Big Dipper's handle is a double star.

The Big Dipper is part of the constellation Ursa Major, which ancient Greeks, Northern Europeans, Micmac Indians of Nova Scotia, and Iroquois Indians along the St. Lawrence River all visualized as a bear. The same pattern in otherwise distinct cultures has stimulated speculation regarding a possible common origin, perhaps transmitted by people crossing the Bering Strait from Asia to North America during the Ice Ages, around 10,000 BC.

In North American Indian mythology, a hunter shot the bear with an arrow, the wounded bear sprayed blood on the hunter, and the hunter shook off the blood, coloring autumn leaves red. In Greek mythology, the

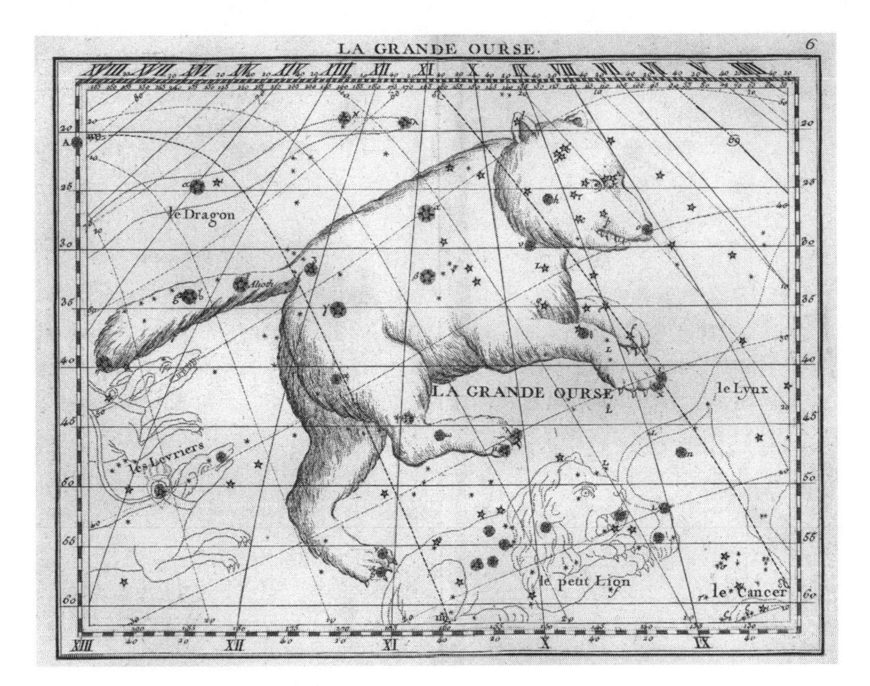

Figure 2.5: Constellation Ursa Major, the Bear. The three stars in the handle of the Big Dipper outline the bear's tail. From a star catalog and atlas by John Flamsteed (1646–1719), an English astronomer and the first Astronomer Royal. (U.S. Naval Observatory Library)

wood nymph Callisto angered Hera, Zeus's wife, by bearing Zeus's son, and Hera changed the girl into a bear. Not knowing that the bear was his mother, the son was about to kill her, when Zeus transformed them both into sky bears and placed them next to each other in the heavens. Substituting Jupiter for Zeus, the Roman poet Ovid (43 BC–AD 17) wrote in his *Metamorphoses*:

> Jupiter snatched them through the air
> In whirlwinds up to heaven and fix'd them there;
> Where the new constellations nightly rise,
> And add a lustre to the northern skies.

Thus Callisto and her son became the constellations Ursa Major and Ursa Minor. The Little Dipper, its handle pointing to Polaris, is an asterism in Ursa Minor.

The Big Dipper may have inspired Mayan mythology. Viewed from Guatemala, the seven stars are overhead at sunset in August, the beginning of the hurricane season, when floods occur, and drop toward the horizon with the rotating of the Earth. Mythical gods created humans, but the first two attempts, one made of mud, and the next of wood, were failures. The third, made from flesh, or maybe from corn dough, was punished with a universal flood. Vukub-Cakix, also called Seven Macaw, had ruled over the world before the flood. Seven Macaw was shot from his perch atop the world tree by Hunahpu, who then, with his twin brother, Xbalanque, resurrected their father, One Hunahpu, the maize god. Hunahpu and Xbalanque became the Sun and the Moon.

The Pleiades cluster of stars may have been recorded as early as 15,000 BC in the Lascaux cave paintings in southwestern France. The Babylonians saw seven warriors and the Greeks saw seven sisters. Most Maya pictured the star group as a rattlesnake, while Maya in the Yucatan saw a rattlesnake's rattle. They used the Pleiades to tell time at night, and their priests carried short sticks with rattlesnakes' tails attached. The Moon was sometimes called the lady in the rattlesnake constellation, presumably referring to passage of the Moon in front of or through the Pleiades. (The Moon as seen from Earth is just large enough to cover the Pleiades.) In Greek mythology, seven sisters were pursued by the hunter Orion, rescued by the gods, changed into doves, at their death transformed into stars, and are now chased again by Orion, this time across the sky.

The French used the word *Pléiade*, the singular of *Pleiades*, first to denote a group of seven tragic poets of ancient Alexandria, and later a group of seven sixteenth-century French poets. The English word *pleiad*, first used in 1839, means a group of seven illustrious or brilliant persons or things. In 1927, seven private, independent, northeastern women's colleges (Barnard, Bryn Mawr, Mount Holyoke, Radcliffe, Smith, Vassar, and Wellesley) dubbed themselves the Seven Sisters, corresponding to the eight Ivy League men's colleges.

A scorpion killed Orion and both were placed in the sky. The constellations are at opposite ends of the sky, Scorpio rising as Orion is setting, so they cannot quarrel with each other. The three stars in a row comprising Orion's belt are among the best-known and easiest-to-find stars in the sky.

The constellation Canis Major, the great dog, follows his master Orion across the sky. Sirius, the Dog Star, is a member of Canis Major and the brightest star in the night sky. Ancient Egyptians and Romans believed that Sirius (Greek: *Sirius* = scorching) added its heat to the summer Sun, creating the extremely hot and sultry dog days of summer, which coincided with the heliacal rising of Sirius. One Greek wrote that the bones of men crumbled under parching Sirius, and that the piercing power and sultry heat of the Sun lessened when the star Sirius spent less time over the heads of men during the day and took a greater share of the night. Another Greek writer urged people to wet their lungs with wine because Sirius was coming round and everything was thirsty under the heat. Two more writers pitched in: one noting that leaves were spreading their overreaching shade against the scorching dog star; the other writing that the star that blazed keenest of all with a searing flame, when he rose with Helios, the Sun, was no longer deceived by the feeble freshness of tree leaves, but with his keen glance pierced their ranks. Some writers noted that Sirius was close pressing upon the Nemean Lion's back (the constellation Leo, which marked midsummer), and that Sirius in autumn, when Mercury and Jupiter grew dim, sharpened yet more his fires. Scorched by Sirius, ancient Minoans raised a great altar to Zeus, who then sent for forty days the Etesian Winds (Greek: *etos* = year, suggesting a regular annual occurrence) to cool the Earth. The American writer Richard Harding Davis (1864–1916) imagined a dog wishing that when the hot days came, people would remember that the days were dog days, and leave a little water outside in a trough, like they did for horses. Summer dog days are often a slow time for stock markets, and poorly performing stocks are derisively called dogs.

The Zodiac

The Babylonians formulated and employed constellations as celestial reference points in the zodiac, an imaginary band in the sky centered on the ecliptic, the path of the Sun across the heavens as seen from the Earth. Later, Babylonians divided the zodiac into twelve equal regions, each identified by a constellation, or sign. The Hindu and the Chinese also divided the zodiac into twelve zones; they could have done so independently of the Babylonians and of each other, guided by the common fact of approximately

Woman with Zodiac

Woodcut, circa 1502, by the German artist Albrecht Dürer. The Earth is in the center of the sphere of the stars. Tilted at an angle of twenty-three-and-a-half degrees to the Earth's equator is the ecliptic, the Sun's apparent path around the sphere of the stars (in a heliocentric model, the intersection of the plane of the Earth's orbit around the Sun with the sphere of the stars). The zodiac is a band centered on the ecliptic and extending above and below the ecliptic, delimiting the observed motions of the Moon and the planets above and below the ecliptic. The zodiac traditionally has been divided into twelve equal sections, each characterized by a particular grouping of stars, called constellations. Dürer was more than a craftsman; he was an artist with a divine gift and he was a cultured intellectual in the age of Renaissance Humanism knowledgeable about mathematics and astronomy.

Figure 2.6: *Woman with Zodiac.* Woodcut, circa 1502, by Albrecht Dürer.

twelve lunar months in a solar year. The Greeks copied the Babylonian zodiac and transmitted it to the Romans.

The Babylonian-Greco-Roman zodiacal signs are Aries (ram), Taurus (bull), Gemini (twins), Cancer (crab), Leo (lion), Virgo (girl), Libra (balance), Scorpio (scorpion), Sagittarius (bow), Capricorn (antelope), Aquarius (pitcher), and Pisces (fish). The Latin word *zodiacus* is derived from the Greek word for circle of animals. The Chinese also have zodiacal animals: rat, ox, tiger, rabbit, dragon, snake, horse, sheep or goat, monkey, rooster or phoenix, dog, and pig. The Chinese associate their zodiacal animals with solar years, in a twelve-year cycle. The year of the rat was AD 2008, the year of the pig 2007, the year of the dog 2006, and so forth. The Maya, whose zodiac had thirteen star groupings, also imagined animals for their constellations. The Roman constellation Aries was an ocelot for the Maya, Gemini a bird, Cancer a frog, Leo a peccary, Libra a bird, Scorpio a scorpion, Sagittarius a fish-snake, Capricorn a bird, Aquarius a bat, and Pisces a skeleton. The Maya's thirteenth zodiacal constellation was the Pleiades. Many cultures, both ancient and modern, have shared the belief that the annually turning zodiac influences events on the Earth, including individual fortunes.

Figure 2.7: *The Northern Hemisphere of the Celestial Globe.* Woodcut, circa 1515, by Albrecht Dürer.

Each zodiacal constellation has its associated myths. Aries is the ram of the Golden Fleece, sought by Jason and the Argonauts. Taurus is the bull that Zeus turned himself into in his seduction of Europa. Cancer is the crab sent by Zeus' jealous wife to nip at Hercules' heels while he battled the Hydra. Of Leo, myths say that the child of Zeus and Selene fell from the Moon to the Earth, where he terrorized the people of Nemea. Hercules, in the first of his twelve labors, killed the Nemean Lion, who was then carried back to the heavens. Shakespeare's Hamlet hoped to be as strong as the muscles (*artire* and *nerve* both mean sinew, or muscle) of the Nemean Lion:

My fate cries out
And makes each petty artire in this body
As hardy as the Nemean lion's nerve.

The Dendera Zodiac

Studies of a zodiac sculpted on the ceiling of a temple at Dendera, Egypt, reveal some of the convolutions and confusions that may arise when one culture is interpreted by another. The Dendera zodiac casts limited light on the culture of ancient Egypt and more light on the culture of those who later studied the temple.

An artist accompanying Napoleon Bonaparte's invasion of Egypt in 1798 brought back to France a sketch of the vault of the heavens shown on the Dendera ceiling. French scientists, distrustful of ancient written texts filtered through the sieve of human culture, were especially fascinated with what they thought was the product of original sensations stimulated by the natural world, and thus free of cultural influences.

Astronomers calculated the date of the zodiac from precession, which is the result of a wobble in the Earth's axis, and which results in the apparent movement of the stars, as seen from the Earth, around the sky in a 26,000 year cycle. It is like a hand on a clock moving around the sky and telling time. The astronomers were confident that the Dendera zodiac depicted summer solstice in the constellation Leo, where it would have been around 2000 BC, even though the beginning point in a circular zodiac is far from obvious and the choice of summer solstice by the Egyptians, while likely, was not certain.

Noah's flood had been assigned a date not much earlier, so the tentative Dendera date conflicted with established religion, and the Vatican soon complained. Threats to religion might not have mattered much in France; a few years earlier a visiting English scientist had noted that all the philosophical persons he had met in Paris were unbelievers in Christianity and even professed atheists. In 1801, however, Napoleon reached an agreement with the pope to reestablish the Catholic Church in France, and newspapers were forbidden to print articles critical of religion. Censorship eased only after the restoration of the monarchy in 1814.

Public interest in the Dendera zodiac reached a new height in the 1820s after a Frenchman using special saws—and when they failed, gunpowder—blasted the zodiac out of its ceiling, brought it to France, and put it on temporary display in the Louvre, originally the royal palace but converted to a museum in 1793. Tremendous public excitement pressured the king into buying the zodiac for France.

A sketch of the Dendera zodiac with surrounding hieroglyphs, which were just beginning to be deciphered, was published in a book. It was now thought that the zodiac might not represent actual observations of stars, but was instead an astrological chart linking celestial events with human destiny; it seemed that the zodiac was not an accurate representation of nature, but was a cultural artifact.

An expedition to Dendera in 1828 produced another surprise: the spaces surrounding the zodiac did not contain the hieroglyphs shown in the published sketch. Evidently an imaginative draftsman had added to the book illustration hieroglyphs from another sketch—so much for linguistic analyses using the spurious hieroglyphs and any date so derived.

The absence of hieroglyphs, however, does suggest a date for the temple ceiling. It may have been completed after the death of Cleopatra's father, in 51 BC, and before 42 BC, when her son by Julius Caesar gained the right, after Caesar was officially made a god, to become king; during the turmoil of the interregnum, there would have been no royal names to inscribe in hieroglyphs around the zodiac.

Figure 2.8: The Dendera Zodiac. Sketch of constellations sculpted in bas-relief, the images raised from the background, on a sandstone slab from the ceiling of the Temple of Hathor at Dendera, on the Nile River near Luxor, Egypt. Four women, one in each corner, helped by falcon-headed spirits hold up the vault of the heavens. From *Description de l'Égypte, ou Recueil des observations et des recherches, qui ont été faites en Égypte pendant l'expédition de l'Armée Française.* C. L. F. Panckoucke, Paris, 1820–1829, volume 4, plate 21. The sandstone slab is now in the Louvre Museum, Paris.

CONCLUSION

Some people find cosmic significance in myths as the psychic residue of past experiences of individuals' ancestors now inherited in the structure of the brain and stirring in the mind primordial images and profound insights. This speculation, however, lacks evidentiary support.

Other people think of myths, when they are bothered to think about them at all, as trivial superstitious fairy tales with no intellectual significance, and whose time has long passed. An important function of myths has always been to characterize the universe in familiar terminology and categories, but science now explains natural phenomena more consistently and thoroughly than myths ever could. Myths can still be employed for teaching moral lessons, but only if they are altered to support changed pedagogical goals. And the provision of entertainment is now better met for many people by movies, television, and video games.

Astronomical myths no longer fulfill important functions they once did, but a residuum survives, woven throughout Western literature, art, and music.

RECOMMENDED READING

Bierhorst, John. *The Mythology of South America* (New York: William Morrow and Company, 1988).

Burland, Cottie. *North American Indian Mythology* (London: Paul Hamlyn, 1965; New York: Tudor Publishing, 1965).

Graves, Robert. *The Greek Myths*, vols. 1 & 2 (Hammondsworth, Penguin Books, 1985–1986).

Kerenyi, C. *The Gods of the Greeks* (New York: Grove Press, 1960).

Ovid. *The Metamorphoses of Ovid, translated and with an introduction by Mary M. Innes* (Hammondsworth: Penguin Books, 1955).

WEB SITES

Allen, Richard Hinckley. *Star Names: Their Lore and Meaning*: http://penelope.uchicago.edu/Thayer/E/Gazetteer/Topics/astronomy/_Texts/secondary/ALLSTA/home.html.

Arnett, Bill. *The Nine Planets: A Multimedia Tour of the Solar System*: http://www.nineplanets.org.

Encyclopedia Mythica: http://www.pantheon.org.

National Gallery of Art. *Quest for Immortality: Treasures of Ancient Egypt*: http://www.nga.gov/exhibitions/2002/egypt/index.shtm.

Out of This World: The Golden Age of the Celestial Atlas. An Exhibition of Rare Books from the Collection of the Linda Hall Library: http://www.lhl.lib.mo.us/events_exhib/exhibit/exhibits/stars/welcome.htm.

Rothman, Susanna. *List of Emblems of Classical Deities in Ancient and Modern Pictorial Arts*: http://homepage.mac.com/cparada/GML/003Signed/SREmblems.html.

Theoi Greek Mythology. *Exploring Greek Mythology in Classical Literature and Art*: http://www.theoi.com.

van Gent, Robert Harry. *Historical Celestial Atlases on the World Wide Web*: http://www.phys.uu.nl/%7Evgent/celestia/celestia.htm.

3

Babylonian Astronomy and Culture

Archaeoastronomers have found astronomical alignments at ancient ruins, but the stones have remained largely silent on why ancient cultures were interested in the heavens. Myths, too, testify to human fascination with celestial phenomena, and reveal more about peoples' conceptions of their place in the universe. Written records further advance our understanding of interrelationships between astronomy and culture. The Babylonians left all three: ruins, myths, and texts.

Mesopotamian astronomical texts from the first and second millennia before Christ reveal a culture concerned with relations between gods, heavenly phenomena, and human society. Celestial divination, or astrology, employing astronomical phenomena to foretell future events, was a prominent part of Mesopotamian cultures, from the earlier Sumerians and Assyrians to the later Babylonians.

From around 500 BC onward, late Babylonian science displays an impressive array of astronomical observations and mathematical techniques for calculating positions of celestial bodies. Indeed, the empirical, quantitative, and predictive nature of Babylonian astronomy, particularly its high level of mathematical procedures, is celebrated by some historians as the beginning of modern science.

MESOPOTAMIAN CIVILIZATION AND WRITING

Mesopotamian civilizations in the valley between the Tigris and Euphrates rivers (modern-day Iraq) were beneficiaries of amazingly fertile soil. The Greek historian Herodotus (ca. 484–425 BC) wrote that blades of wheat and barley were three inches wide, and harvests were two hundred times what was normal elsewhere. Sumerians controlled the lower part of the Tigris-Euphrates valley from long before 3000 to around 2000 BC; Hammurabi (ca. 1792–1750 BC), known for his code of laws and also for not deifying himself, united much of the country; Assyrians ruled from the ninth century BC until a revolution and the destruction of Nineveh in 612; the Persians captured Babylon in 539; and in 330 Alexander the Great conquered the Persian Empire; his followers, the Seleucids, ruled Babylon for the next three centuries.

For much of this time, scribes pressed sharpened sticks into soft clay tablets about the size of a hand, leaving wedge-shaped, cuneiform (Latin: *cuneus* = wedge and *forma* = shape) signs representing speech sounds, words, and even concepts. Amazingly durable if kept dry, these clay tablets now provide a glimpse into life in Mesopotamia from thousands of years ago.

CELESTIAL DIVINATION

The Roman writer Cicero thought that in all ancient cultures, whether refined and learned or savage and ignorant, people believed that signs are given of future events and that some persons can recognize the signs. He particularly singled out the Assyrians, who had inhabited vast plains in northern Mesopotamia with open and unobstructed views of the heavens, and thus more easily observed movements of the stars. In the south, the Babylonians, too, according to Cicero, had from long-continuing observations of the constellations perfected a science enabling them to foretell any person's fate.

Mesopotamians believed that gods had created the universe and that all the phenomena of the physical world, including planetary and lunar positions, equinox and solstice dates, and eclipses, were divinely ordered and ordained. Furthermore, they recognized that the phenomena occurred with observable regularity. Omen texts stated that if one thing occurred, then another would follow—or, less specifically, that whatever the result, it would be favorable or unfavorable.

One omen text simply states what would happen if the king took part in ceremonies in each of the twelve months of the year. Another warns what to expect depending where clouds are in relation to the Sun. Other omen texts have ties to different celestial phenomena, including lunar eclipses: an eclipse in the evening warns to watch for plague; an eclipse in the morning indicates curing of illnesses; an eclipse in the south foretells the downfall of

Figure 3.1: Astronomer's Manual. This tablet instructs astronomers how to go about their business. They should study the length of the year and look in tablets for the times when stars first appear, are visible, and then disappear. They should observe when the Moon rises and first appears each month, and when the Pleiades is opposite the Moon. Finally, they should do perfectly whatever they are doing. Tablet WA K2847, British Museum.

the Subarians and the Akkad; an eclipse in the west foretells the downfall of the Amorites; an eclipse in the north foretells the downfall of the Akkadians; and an eclipse in the east foretells the downfall of the Subarians (again). Planetary appearances were also omens: if Jupiter remained in the sky in the morning, enemy kings would become reconciled; if at Venus's rising, the red star (Mars) was also nearby in the sky, the king's son would seize the throne.

In modern Western culture, astronomy is a widely recognized and highly respected science. Astrology, claiming divination by the positions of the planets, Sun, and Moon, is not a science, at least not viewed as such in the scientific community. A poll in 2007 showed that 25 percent of Americans and 52 percent of Europeans believed that astrology is scientific, but astrology is still classified as a pseudo science by the educated elite. Major newspapers print astrology columns, but label them as entertainment. Furthermore, astronomy and astrology are practiced by different groups of people with different training and patrons. In ancient Mesopotamian culture, however, astronomy and astrology were not separate disciplines. Beginning around the fifth century BC, personal horoscopes expanded the realm of celestial divination beyond king and state. Some horoscopes relied on celestial phenomena observable at the time of birth (nativity omens): if a child was born when Venus came forth and Saturn set, his oldest son would die; if a child was born when Venus came forth and Jupiter set, his wife would be stronger than he was. Other horoscopes were based on more detailed and precise knowledge of celestial phenomena: the positions of the Sun, Moon, and planets in the zodiac on the hour and day the person was born. Born in Taurus, a man would be distinguished, his sons and daughters would return, and he would see gain.

MATHEMATICAL ASTRONOMY

It is tempting to think in modern terms and imagine a scientific discipline of systematic, mathematical, computational, predictive astronomy emerging out of and separating from Mesopotamian celestial divination. With the advent of personal horoscopes, observation of nightly planetary positions would no longer have sufficed for celestial divination because many births took place during the day. Nor is every planet necessarily visible above the horizon during nocturnal births. It would have become increasingly necessary to compute astronomical positions, especially those upon which horoscopes depended. Rudimentary mathematical computational methods in Babylonian astronomy predate increased interest in personal horoscopes during the last half of the first millennium BC. Yet the accompanying need for astronomical computations suggests that the approximately simultaneous development and flourishing of horoscopes and mathematical astronomy was more than coincidence. They were related and complementary aspects of ancient Mesopotamian culture.

Mesopotamians had long sought to recognize indications of future events in celestial phenomena. Naturally enough, and no doubt encouraged by the noted regularity of some celestial motions, Mesopotamians also sought to predict occurrences of indicative celestial phenomena. A tablet from the second millennium BC reveals both knowledge of future planetary appearances and a corresponding foretelling of events. If Venus disappeared on the seventh day of the month of Kislimu and remained invisible for three months, and then rose in the west on the eighth day of Addaru, king would send messages of hostility to king.

One of the most remarkable discoveries by Babylonian astronomers is the Saros cycle, the time period of 6,585 days (18+ years) for the nearly exact recurrence of three different celestial alignments, all necessary for the recurrence of an eclipse. Had a total solar eclipse been observed on May 18, 603 BC, an astronomer could have confidently predicted a similar eclipse on May 28, 585 BC.

Historical Eclipses

Reports of ancient eclipse observations hold out hope of dating specific events. Herodotus wrote that after five years of indecisive war between the Medes and the Lydians, day was suddenly turned into night while they were engaged in a battle. Furthermore, the Greek philosopher Thales of Miletus (ca. 625–547 BC) supposedly had fixed the date for the solar eclipse within the limits of the year in which it took place. According to modern calculations, a total solar eclipse was observable in what is now northern Turkey on May 28, 585 BC.

Other memorable eclipses include an unexpected lunar eclipse in 413 BC, which delayed the departure of an army from Athens, and a total lunar eclipse in AD 549, which persuaded two Teutonic armies about to fight each other to run away instead, leaving their leaders no alternative but to conclude an armistice.

Christopher Columbus (1451–1506) is also linked with an eclipse. To persuade recalcitrant Indians (he thought he had arrived in the East Indies) that God was angry because they had failed to keep Columbus and his crew amply supplied with food, Columbus warned the Indians that that very night they would see a token of the punishment God was going to visit upon them. The Indians' merriment turned to concern as the Moon rose inflamed with wrath, and they hurried off to find supplies. Columbus retired to his cabin, not to pray to God to forgive the Indians, as he had assured them he would, but to calculate from his observations of the lunar eclipse the longitude of his ship. He had calculated longitude that way before and had known in advance that an eclipse would occur that evening.

Some astronomical tablets contain only observations, without accompanying prognostications. One tablet reports monthly observations; another provides observations of the planet Mercury; and yet another lists lunar eclipses. A much more substantial tablet lists stars and constellations along the paths of the particular gods, reports when stars and constellations became visible, and ends with a list of gods (individual stars and constellations) on the path of the Moon.

By around 1000 BC, Mesopotamian astronomers had devised a calendar with the year divided into twelve months, each month marked by the heliacal risings (becoming visible just before dawn) of three stars. The paths of Anu, Ea, and Enlil were in the northern sky, in a central band lying around the celestial equator, and in the southern sky.

Only later, around 500 BC, did Babylonian astronomers begin to use for celestial references the ecliptic (the intersection of the plane of the Earth's annual path around the Sun with the celestial sphere) rather than paths of gods. Objects far from the ecliptic, however, might still be referenced to godly paths. What is now known to have been Halley's comet was observed in 164 BC in the east in the path of Anu in the area of the Pleiades and Taurus, then to the west, and finally in the path of Ea.

Over time, Babylonian astronomy became more mathematical and scientific. During the last three centuries BC, the Babylonians devised arithmetic progressions precisely describing motions of celestial bodies. An ephemeris, a tabular statement of the positions of a celestial body at regular intervals, for the years 133–132 BC enables calculations of the position of the Sun.

Babylonian Ephemeris for the Sun, 133–132 BC

XII	28,55,57,58	22,8,18,16	Aries
I	28,37,57,58	20,46,16,14	Taurus
II	28,19,57,58	19,6,14,12	Gemini
III	28,19,21,22	17,25,35,34	Cancer
IV	28,37,21,22	16,2,56,56	Leo
V	28,55,21,22	14,58,18,18	Virgo
VI	29,13,21,22	14,11,39,40	Libra
VII	29,31,21,22	13,42,1,2	Scorpio
VIII	29,49,21,22	13,32,22,24	Sagittarius
IX	29,56,36,38	13,28,59,2	Capricorn
X	29,38,36,38	13,7,35,40	Aquarius
XI	29,20,36,38	12,28,12,18	Pisces
XII	29,2,36,38	11,30,48,56	Aries

From Otto Neugebauer, *Exact Sciences in Antiquity*, 2nd ed. (Providence, RI: Brown University Press, 1957; New York: Dover Publications, 1969), p. 110.

The Babylonians used a sexagesimal number system, based on the number 60. Positions were reported, for example, as 28, 55, 57, 58, with each succeeding unit representing so many 60ths of the preceding unit—thus 28 degrees, 55 minutes, 57 seconds, and so forth. Western civilization still divides hours into 60 minutes and minutes into 60 seconds.

The first column of the ephemeris lists the months, from I through XII.

The second column lists how far the Sun moves in each month. During month I, the Sun moves 28 degrees, 37 minutes, 57 seconds, and 58/60ths of a second (line 2, column 2). Add this movement to the position of the Sun at the end of the previous month (line 1, column 3), and the result is the position of the Sun at the end of month I (line 2, column 3).

The third column gives the position of the Sun. It is obtained by adding to the initial position the amount of motion during the month, listed in the second column. Adding the top line of the third column and the second line of the second column produces the second line in the third column—after subtracting 30 degrees and listing the position 30 degrees ahead in the next segment of the zodiac.

The fourth column is the house of the zodiac in which the Sun resides that month. Each house of the zodiac occupies 30 degrees, one-twelfth of the 360 degrees of a complete circle.

This ephemeris shows a decrease of 18 minutes in the distance traveled by the Sun from one month to the next for the first two months of the year, then an increase of 18 minutes for each of months 3 through 8, and then a decrease in distance of 18 minutes in each consecutive month over the last four months. Similar mathematical models were used for the Moon, its velocity changing with time, increasing in steps of a day each from minimum to maximum speed, and later decreasing.

For the Sun, within each of the three groups of months (decreasing distance, then increasing, then decreasing again), the last two sets of numbers (seconds and sixtieths of a second) in each line of the column telling how far the Sun moved that month are unchanged. Obviously, the numbers do not represent actual observations; they must have been calculated by adding or subtracting a fixed number of minutes (18 in this case) to or from the distance traveled by the Sun in the preceding month. The observational precision implied, to sixtieths of a second, is also spurious. Babylonian observations, made without benefit of telescopes, would have been accurate only to tens of minutes of arc, at best.

The decreasing, increasing, and then decreasing again sequence of numbers for the distance traveled by the Sun from month to month in Babylonian mathematical models now is called a zigzag function, after its appearance on a graph. Babylonian astronomers also used a system in which the solar velocity remained constant for several months, after which the Sun proceeded with a different constant speed for several more months, before reverting to the initial velocity and remaining at that speed for several more months. Graphed, the motion looks like a series of steps up and down, and accordingly is called a step function. Zigzag and step are useful labels, but potentially misleading, because the Babylonians are not known to have used graphs.

CULTURAL EXCHANGES

The precision of mathematics occasionally reveals evidence of exchanges between cultures. The Greek astronomer Ptolemy (ca. AD 90–168), who did for astronomy what Euclid had done for geometry and earned a reputation as the greatest astronomer of the ancient world, noted that there are 4,267 synodic months in 126,007 days and 1 hour. Then, to obtain the value of the mean synodic month, Ptolemy purportedly divided 126,007 days and 1 hour by 4,267. He reported that the mean synodic month was 29, 31, 50, 8, 20 (29 plus 31/60 plus 50/3600 etc.) days, which is also the Babylonian value for the mean synodic month. The correct answer to Ptolemy's purported division, however, is 29, 31, 50, 8, 9. Obviously, Ptolemy did not do the calculation; instead, he must have borrowed the Babylonian value for the mean synodic month. This limited exchange of astronomical information between the Babylonians and the Greeks hints at a wider range of mutual influences between Oriental and Hellenistic cultures.

WAS IT SCIENCE?

Some historians have characterized Babylonian astronomical texts of the Seleucid period as scientific because everything was eliminated from the astronomy except observations and mathematical procedures, leaving no role for speculative hypotheses of a theoretical nature. These same historians then proceed to celebrate the high level of mathematics as the first appearance of modern science.

Babylonian astronomy, however, differed significantly from modern science. Historians concerned with underlying motives of science and more appreciative of scientific theories dismiss Babylonian astronomy as little more than a set of mechanical procedures with no more intellectual content than recipes in a cookbook. Babylonians studied how the celestial motions went, but not why. No suggestion is found in cuneiform texts that the Babylonians considered that the celestial motions they were observing might be controlled by laws of nature. Also, later Greek concepts of geometrical models and continuous motion along a circular path are absent from Babylonian astronomy, whose primary subjects were intermittent celestial positions and period relations. Nor did Babylonians think of their mathematical structures as an image of the world; there was no commitment to a cosmological framework. Despite all their sophisticated mathematical astronomy, the Babylonians' universe remained more a chaos than a cosmos; it lacked a rational and scientific understanding. Babylonians did not attempt to develop a single comprehensive mathematical scheme encompassing all their data. In its proliferation of different mathematical models for different celestial bodies, rather than insistence on a single model encompassing all the phenomena, Babylonian astronomy reveals the existence of cultural values very different from current ones. Explanatory expectations can differ from culture to culture, and Babylonians did not seek the types of explanation characteristic of modern science. Many historians of science look instead to the Greeks for the birth of modern science.

RECOMMENDED READING

Chiera, Edward, with George G. Cameron, ed. *They Wrote on Clay: The Babylonian Tablets Speak Today* (Chicago: University of Chicago Press, 1938, 1966).

Neugebauer, Otto. *Exact Sciences in Antiquity*, 2nd ed. (Providence, Rhode Island: Brown University Press, 1957; New York: Dover Publications, 1969).

———. *Astronomy and History: Selected Essays* (New York: Springer-Verlag, 1983).

Rochberg, Francesca. *The Heavenly Writing: Divination, Horoscopy, and Astronomy in Mesopotamian Culture* (Cambridge: Cambridge University Press, 2004).

Swerdlow, Noel. *The Babylonian Theory of the Planets* (Princeton: Princeton University Press, 1998).

WEB SITE

"Astronomers of Babylon," in *Livius: Articles on Ancient History*: http://www.livius.org.

4

π in the Sky

More or less continuously from the Greeks in the fourth century BC to the European Renaissance in the sixteenth and seventeenth centuries AD, a common cultural value persisted across a variety of geographical settings, empires, religions, and civilizations. Many astronomers—pagan, Islamic, and Christian alike—believed that the observed irregular motions of the Sun, Moon, and planets were combinations of circular motions with constant, uniform speeds, and they accepted as their primary task the "saving of the appearances" with systems of uniform circular motions. This remarkably persistent cultural exaltation and elevation of things circular significantly shaped expectations and requirements for an aesthetically satisfying understanding of nature.

The enterprise and the geometrical models produced have been characterized as *π in the sky*, a piquant and provocative word play on the fact that the circumference of a circle is equal to its diameter times the number π, and π is pronounced *pie*. The homophonous expression *pie in the sky* comes from the Swedish-American labor agitator Joe Hill's 1911 song *The Preacher and the Slave*, a parody of the Salvation Army hymn *In the Sweet Bye and Bye*. Hill (1879-1915) wrote:

> You will eat, bye and bye
> In that glorious land above the sky;
> Work and pray, live on hay,
> You'll get pie in the sky when you die.

EARLY GREEK PHILOSOPHY

In contrast to the astrological superstition permeating Babylonian astronomy, early Greek philosophical thought has been hailed as a miracle of rationalism. Supposedly it was based on evidence and buttressed by reason, unargued fables were replaced by argued theory, dogma gave way to reason, celestial phenomena were explained in natural terms rather than attributed to supernatural or divine intervention, and an entire mythological scaffolding of earlier prescientific thought was removed at a single stroke.

No matter that no Greek astronomical or cosmological theory has survived the test of time, nor that Greek astronomy lacked for centuries the systematic mathematical science found in late Babylonian astronomy. The early Greeks were incurably vague, innocent of the delights of measurement, quantification, and the experimental method. There was no Ancient Greek Royal Scientific Society—not that ancient people lacking knowledge of the future could have aspired to become modern scientists—nor was there much lasting contribution to the sum of scientific knowledge. Early Greek philosophy was a speculative enterprise without a scientific future, a philosophical sideline with little impact on the development of observational science or mathematical astronomy.

Eventually, however, Greek philosophers developed a quantitative geometrical astronomy unrivaled in antiquity, surpassing all science that had gone before, not itself surpassed for fourteen centuries, and inextricably intertwined with different cultures in different times and places. Scientific attitudes and modes of thought impacted wider cultural concerns, just as social and political change preceded and shaped consequent changes in astronomy.

New Stone Age or Neolithic culture, characterized by agriculture and the domestication of animals, spread to Greece during the fourth millennium BC. Invaders from the north developed the Mycenaean civilization around 2000 BC. At the same time, Minoans on Crete acquired both trade goods and ideas from Egyptian and Asian civilizations. Minoan civilization collapsed around 1400 BC, cutting off Greece from its Near-Eastern links. Trade, technology, and culture were swept away in the ensuing Dark Age, from about 1200 to 800 BC. When revival occurred, it was without the Oriental influences of the earlier period.

The political base in Greece broadened and the city-state developed as a form of government. Political debate in small-scale, face-to-face social settings presumably carried over to philosophical thought and pushed astronomy toward rationalism.

Conversely, attitudes and ideas of the new astronomy conflicted with older religious, ideological, and moral traditions, saving Greece from political and intellectual petrification. Influenced by the fall of a meteorite from space, Anaxagoras (ca. 500–428 BC) taught that the Sun and the stars were red-hot stones carried around with the rotation of the ether. For reducing the god Helios to a lifeless lump of stone and threatening the popular belief that celestial phenomena were controlled by gods, Anaxagoras was exiled from Athens.

Another threat to the gods came from Xenophanes (570–480 BC), who reduced the multiple gods of mythology to meteorological phenomena, particularly clouds. In Aristophanes' play *Clouds*, first performed in Athens in 423 BC, astronomers are depicted looking for onions on the ground, bent double, their buttocks looking at the heavens, studying astronomy on their own account. A pompous philosopher, Socrates, floats by in a basket, traversing the air and contemplating the Sun. He says there are no gods. Rain comes from clouds, not from nymphs pouring pitchers of water, nor from Zeus pissing into a sieve; and the roll of thunder is the sound of clouds bumping against one another, not gods defecating. At the end of *Clouds*, one of the characters moans:

> Oh! What madness! I had lost my reason when I threw over the gods through Socrates' seductive phrases.
> Let us burn down the home of those chatterers ... and may the house fall in upon them.
> Ah! You insulted the gods! You studied the face of the Moon!
> Chase them, strike and beat them down! Forward! They have richly deserved their fate—above all, by reason of their blasphemies.

In 399 BC, Socrates was found guilty of impiety and put to death.

The Ionians

New Greek cities were established around the Mediterranean Sea, and with them came new trading opportunities and increased prosperity. In the region of Ionia (on the coast of Asia Minor, now Turkey), particularly in the city-state of Miletus, the richest in the Greek world, a new approach to the understanding of nature was formulated: that matter or principles in the form of matter were the principle of all things.

Thales argued that everything was formed out of water. Very likely this is an instance of philosophy influenced by practical life, techniques, and economics, in that water was so important in agriculture. Water was also central in Babylonian and Egyptian accounts of the creation of the world. However vague and unsusceptible to verification Thales' theory may have been, it was the first general explanation of nature not to invoke aid from an outside power.

Thales' pupil Anaximander (610–546 BC) agreed that there was one basic principle of nature, but the first thing was not any specific substance; rather, it was something indefinite. This concept was an advance toward a more abstract conception of nature: no longer was the underlying substance of the material world a visible, tangible state of matter, such as water, but it was the lowest common denominator of all sensible things, arrived at by a process of abstraction. Anaximander thought that celestial bodies were wheels of air filled with fire escaping through holes. His pupil Anaximenes (ca. 585–525 BC) asserted that the basic principle was air and the infinite.

Miletus was destroyed by the Persians in 494 BC, but not before its tradition of searching for a basic principle of nature had spread to nearby cities. Heraclitus (late sixth century BC) chose fire, the active agent producing change in many technical and natural processes, as the first principle. Xenophanes proposed that everything was composed of two elements, water and earth.

Xenophanes later traveled to Sicily. Here Empedocles (ca. 490–430 BC) subsequently proposed that there were four basic elements (earth, water, fire, and air), and that different mixtures of these elements produced different substances. Love and strife, the two dynamic first principles, united and separated substances. The heaven was air fixed in crystalline form by fire, and the Sun, planets, and stars were composed of fiery matter. The Moon had its light from the Sun, and a solar eclipse was caused by the Moon passing in front of the Sun.

In *Empedocles on Etna*, based on the legend that Empedocles jumped into the active volcano Mount Etna to persuade people that his body had vanished and he had become an immortal god, the English poet Matthew Arnold (1822–1888) wrote:

> To the elements it came from
> Everything will return.
> Our bodies to earth,
> Our blood to water,
> Heat to fire,
> Breath to air.

The Pythagoreans

A second major tradition in early Greek philosophy centered on Pythagoras (ca. 580–500 BC), who was born on the island of Samos, near Miletus. He traveled in Asia, Egypt, and Greece before settling in southern Italy, where he founded a religious fraternity devoted to mathematics and numbers as the principle of all things.

Contempt for manual labor and technology likely had increased along with slavery in ancient Greece, and might have encouraged Pythagorean philosophers to focus their attention on the secret constitution of things. These were revealed not to tinkerers who manipulated nature, but to thinkers.

From their observation that attributes and ratios of musical scales are expressible in numbers, Pythagoreans concluded that the heaven was composed of musical scales and numbers, and the planets' movements produced harmonious music. Because the number ten was perfect and comprised the whole nature of numbers, there must be ten celestial bodies. Adding up known bodies, however, the Pythagoreans only counted nine: Sun, Mercury, Venus, Earth, Moon, Mars, Jupiter, Saturn, and the sphere of the stars. So they imagined the existence of a tenth body, a counter-Earth, moving around the central fire, the Sun, and on the other side of it from the Earth.

Pythagoreans also attempted to understand qualities, including justice and goodness, in terms of numerical ratios, and their mathematics and astronomy is sometimes characterized as primarily a religious exercise. Pythagoras was not the leader of a scientific research team or an educational establishment, but rather the guru of an Indian ashram or a combination of Einstein and Mary Baker Eddy (the founder of the Church of Christ, Scientist). Pythagoreanism was a spiritual revival encouraged by the menace of Persian armies.

Belief in and search for numerical relationships among phenomena of nature has been important throughout much of Western civilization, and remains so today. The astronomer Johannes Kepler (1571–1630) spent much of his working life searching for the numerical ratios that his cultural background convinced him existed in the arrangement of the planetary system. On a Pythagorean foundation of fantasy, Kepler built, by equally unsound reasoning, the solid edifice of modern astronomy. Also, Nicholas Copernicus (1473-1543) was encouraged by Pythagorean philosophy to displace the Earth from the center of the universe and put it in movement around the Sun.

UNIFORM CIRCULAR MOTION

Even more influential than Pythagorean thought for the course of astronomical and cultural development over the next two thousand years was the Greek belief that the motions of the Sun, Moon, and planets around the Earth are actually combinations of circular motions with constant, uniform speeds. Astronomers, however, found that the observed motions of the Sun, Moon, and planets were neither circular nor uniform—hence the need to reconcile apparent motions determined by observation with real motions revealed by reason. To save the appearances with a system of uniform circular motions became the primary goal for astronomers.

Plato

The overall goal of Greek astronomy is generally attributed to Plato (ca. 429–347 BC), although no explicit statement about saving the appearances is found in his surviving writings. The concept is plausible in the context of his philosophy, however, which, in turn, is understandable in the context of his life. Plato's philosophy and its implications for the study of astronomy were a response to the time of troubles in which he found himself.

In 479 BC, a year after the Persians captured and burned Athens, thirty-one Greek city-states defeated the Persians in decisive land and sea battles, capping twenty years of struggle to stop the westward expansion of the Persian Empire and marking the beginning of Greece's Golden Age. Increasingly, Athens dominated, and tribute poured in from other city-states, giving support to Athenian writers and artists. Led by Sparta, several Greek city-states revolted against Athenian rule, setting off the Peloponnesian War of 431–404 BC, which ended with the surrender of Athens. During the turmoil, Socrates encouraged the youth of Athens to question every moral precept handed down to them, and also to question the government and its actions.

Plato was of an age to enter public life about the time of the defeat of Athens, but he judged that the new ruling tyrants were not leading Athens out of the unjust life and establishing her in the path of justice. A year later, a democratic faction drove out the tyrants, and Plato again considered entering politics; but the new democracy persecuted Socrates, Plato's close friend and teacher.

Reacting to fluctuating moral values of his time, Plato searched for unchanging standards. The changing, visible world was without permanent values. Thus Plato turned to the world of ideas, where he hoped to find the real and unchanging standards so sadly absent in his world of experience.

In his allegory of the cave, Plato explained that the prison of the cave corresponds to the part of the world revealed by the sense of sight. Men chained by their necks so they could see only what is in front of them, and seeing only shadows of puppets thrown by fire light on a cave wall, would recognize as reality nothing but the shadows of artificial objects. Escape from the cave corresponds to the use of intelligence to reach the real world of knowledge. If one of the prisoners were freed and forced to stand up and walk toward the cave entrance, his movements would be painful and his eyes would be dazzled by sunlight. Told that what he had formerly seen was meaningless illusion, and that he is now nearer reality and enjoying a truer view, would he believe it? And if he returned to the cave, would his fellow prisoners believe his account of what he had seen outside the cave? Plato concluded, pessimistically, that if the prisoners could lay hands on a man trying to set them free and haul them painfully up, they would kill him—just as, Plato might have added, they had killed Socrates.

Plato's concept of reality is plausibly illustrated with a simple example. Think of a circle, and draw a circle. Which is real? The circle drawn on paper is an imperfect representation in the visible world of a perfect circle, which exists only in the mind, only in the world of thought. Similarly, at the end of the eighteenth century AD, when the German poet and scientist Johann Wolfgang von Goethe (1749–1832) created a symbolic plant with a few strokes of his pen, Friedrich von Schiller (1759–1805), a fellow poet, exclaimed that it was not an experience but an idea. As had Plato, Goethe saw with the eyes of his soul.

Plato's Pervasive Persuasive Philosophy

Plato's philosophy has affected many human intellectual endeavors over many centuries. For example, the Romanian sculptor Constantin Brancusi (1876–1957) consciously and explicitly tried to make his art a working philosophy of Plato. What is real is not the external form but the essence of things, and it is impossible for anyone to express anything essentially real by imitating its exterior surface. Brancusi looked beneath the surface of human experience for a deeper and truer reality. He abandoned details, which expressed little, and kept only essential anatomical elements. His sculptural kissing couple represents an ideal, not just a particular pair of people in love, but all the pairs that have lived and loved on Earth.

Similarly, poets should aspire to stimulate the imagination not with specific details of the appearances of things, but with words that suggest vast ideas. And landscape painters should paint generalizations, not nature's littlenesses. The British painter J. M. W. Turner (1775–1851) was impressed with the power of sunlight and constantly hymned the omnipotent energy and saving grace of the Sun. Works of art and astronomical theories lift up the soul and enable it to participate in the splendors of the divinity.

The popular movie *The Matrix*, released in 1999, is a modern variation on Plato's cave. People are not living in the world they perceive; rather, they lie comatose in cocoons, stacked in incubators, clear pods piled high in high towers, their brains penetrated by cables delivering an interactive virtual-reality program, its simulation mistaken for reality. Another cinematic variation on Plato's cave is the 1998 movie *The Truman Show*, based on Philip K. Dick's 1959 novel *Time Out of Joint*, and also reminiscent of the 1960s cult television series *The Prisoner*. Truman lives, unknowingly, in a continuously running reality televised soap opera, and all the other people, including the woman he believes to be his wife, are actors in what he mistakes for a real town.

Taking up the discussion of astronomy in his book the *Republic*, Plato mentioned its utilitarian benefits in agriculture, in navigation, and in war. The true utility of the regimen of study prescribed in the *Republic*, however, was saving souls.

An obvious way of doing astronomy was to observe motions of celestial objects. But only a discipline dealing with unseen reality, forcing the mind to look away from the world of experience and observation, would lead the mind upward.

Intricate traceries in the sky were the loveliest and most perfect of material things, but they were still part of the visible world, and therefore fell short of the true realities: velocities in the world of pure number and perfect geometrical figures carrying around celestial bodies. Framed by their artificer with the highest perfection possible, they were to be conceived by reason and thought, not seen by the eye. Astronomers should proceed as was done in geometry, by means of problems, and leave the starry heavens alone.

This admonishment to leave the starry heavens alone is anti-empirical, and it could easily have lead to a purely speculative study of bodies in motion with no connection to the world of observations. Such an outcome would have been even more likely if the original Greek admonishment translated here as *let alone* or *leave* was understood in the stronger sense of *dismiss* or *abandon*.

Even if an astronomer were to study the heavens, he was not expected to develop just any theory that fit his observations. He was expected to fit the observations into a preconceived geometrical pattern. Plato believed that his circles in the sky, the product of logical inference and analysis, were more certain than fallible observations, which were all too likely to be proved false by subsequent improvements in observational capability.

Whether Plato's vision for astronomy would die in infancy or grow in strength would depend upon the support it received. Greek society encouraged playwrights when the citizens of Athens paid to see the productions of Aeschylus, Aristophanes, Euripides, and Sophocles, and prizes were awarded at festivals to poets and musicians.

No city held geometry in high regard, however, and inquiries into this subject languished, so Plato lamented. He taught his ideas to pupils at the Academy, possibly the world's first university, which he founded in Athens around 380 BC, and which survived until AD 529. Plato believed that mathematics provided the finest training for the mind, and above the door of the Academy was written *Let No One Unversed in Geometry Enter Here.*

Concentric Spheres

Around 385 BC a poor young man traveling as a physician's assistant arrived in Athens. Eudoxus (ca. 410–350 BC) may have studied with Plato then, before visiting Egypt, where he became familiar with priests' astronomical observations. Later, back in Greece, Eudoxus established his own school. He assigned three spheres each for the Sun and the Moon, and four spheres for each of the planets. All the spheres were centered on the Earth, and all rotated with constant speeds. Eudoxus's greatest triumph was to devise a system of spheres rotating with uniform speeds that reproduced the appearance of retrograde motion, when a planet seemingly ceases its forward movement relative to the stars, turns back temporarily, retraces a small portion of its path, and then changes direction again to resume its voyage around the heavens.

Eudoxus was seemingly content to reproduce retrograde motion only qualitatively, not fully quantitatively. Determining the path traced by a point on the innermost of four rotating spheres was so stupendous and absorbing a feat that it might well have provided its own justification. Nor would lack of descriptive and predictive accuracy necessarily have weighed heavily against a theory intended primarily to give conceptual unity to the celestial motions, to demonstrate in a general way that a small number of principles could account for a large number of phenomena.

There is one phenomenon, however, for which Eudoxean systems cannot account: the planets move at different times closer to and farther from the Earth. No system of concentric spheres can produce changes in distances from the center of the spheres, and soon Eudoxus's theory was abandoned.

Some historians and philosophers of science believe that Eudoxus intended his system of planetary motions as purely a calculating device. In this instrumentalist view, scientific theories are merely fictive; it is enough that they yield predictions corresponding to observations. Because Eudoxus did not take into account changing distances, he must have been an instrumentalist, so the argument goes.

Realists, on the other hand, believe that scientific theories are descriptions of reality, and that Eudoxus believed he was describing bodies and movements that actually were accomplished. Because his system could have been less complex had it been intended merely as a computational device, he must have been a realist. Otherwise, why would he have gone to all the trouble and work?

Eudoxus' system for saving the appearances was far more than an instrumentalist's calculating device, but also less than a realist's empirical reality. Eudoxus was infatuated with Plato's vision of true realities conceived by reason and thought, rather than seen by the eye. These realities were velocities in the world of pure number and perfect geometrical figures carrying around celestial bodies.

Nearly 2,300 years later, Albert Einstein revealed a similar aesthetic faith when he said that the chief significance of the general theory of relativity lay not in the fact that it predicted a few minute observable facts, but rather in the simplicity of its foundation and in its logical consistency. Perhaps Eudoxus believed his vision of the universe so clear and penetrating an understanding that observation was rightly consigned to a minor role. And if he did, can modern scientists following Einstein reject Plato's and Eudoxus's philosophy?

Eccentrics and Epicycles

Not long after Plato and Eudoxus flourished in Athens, the center of scientific activity in the Greek intellectual world shifted to Alexandria, a port city established by Alexander the Great (356–323 BC) on Egypt's Nile River in 332. Around 290, Ptolemy I founded the Museum, home to a hundred scholars subsidized by the government, with lectures, and with specimens of plants and animals collected for study. Not to be outdone, Ptolemy II established the Library. Its famous collection of as many as half a million books was obtained by purchasing private libraries, including possibly Aristotle's. Astronomical instruments were constructed for use at the Library, and the matching of theory with observation was undertaken on a systematic and sustained basis. In ancient Alexandria, much prestige was attached to scholarship and scientific research, and the Ptolemies sought thus to enhance their reputations.

A preliminary geometrical proposition producing retrograde motions of planets was demonstrated by several mathematicians in Alexandria, including Apollonius (ca. 262–190 BC). The same old assumption continued to govern new theories: the motions of the planets were regular and circular by nature. By regular, or uniform, the Greeks meant that the speeds of motion were constant—that is, straight lines revolving the planets on their circles cut off in equal times on all circumferences equal angles at the centers of each. All apparent irregularities then must result from the positions and arrangements of the circles.

Irregularities could be accounted for by two simple hypotheses, the eccentric and the epicycle. In the former, planets made their regular movements along circles not concentric (but eccentric) with the cosmos; in the latter, planets made their regular movements along small circles (epicycles), which were carried around at constant speeds by larger, concentric circles (deferents). According to either hypothesis, it was possible for the planets, as seen from the Earth at the center of the cosmos, seemingly to pass in equal periods of time through unequal arcs. No evidence suggests that Apollonius tried to quantify the models. He may have been perfectly content with his brilliant qualitative explanations, never imagining that the models could take on quantitative, predictive power.

The eccentric and epicycle hypotheses are mathematically equivalent: identical phenomena follow from either hypothesis. On what basis, then, could an astronomer choose between them? Hipparchus (ca. 190–120 BC) preferred the epicycle theory because it better conformed to the nature of things. The epicycle was the circle that a planet traced out as it moved, conforming to nature, as opposed to the eccentric, which was traced out only accidentally. Another Greek astronomer, Claudius Ptolemaeus (ca. AD 90–168), known as Ptolemy but not to be confused with Egyptian pharaohs also named Ptolemy, made the opposite choice. To him the eccentric hypothesis seemed more probable because it was simpler, performed by one motion rather than two.

Simplicity, like beauty, often resides in the eye of the beholder. Present day scientists find neither the eccentric nor the epicycle hypothesis mathematically any simpler than the other, thus suggesting other human values and cultural considerations were involved in the choice of astronomical models.

Quantitative Astronomy

From Apollonius, the history of Greek geometrical astronomy moves on to Hipparchus, who was credited by the astronomer Ptolemy with developing a quantitative solar model. It was based largely on conjecture, however, because Hipparchus knew of very few observations of stars before his own time. According to Ptolemy, Hipparchus did not give the principle of the hypotheses of the planets because he had found it too difficult to reproduce retrograde motion quantitatively, either by eccentric circles, by circles concentric with the ecliptic but bearing epicycles, or even by combinations of both models.

Three centuries later, Ptolemy would possess many more observations extending back in time, particularly Hipparchus's observations. With these observations, Ptolemy would correct Hipparchus's solar model and also correct and develop further Hipparchus's lunar theory.

An Ancient Star Catalog

Hipparchus's star catalog long was thought lost, but in 2005 what is probably a pictorial presentation of it, in plain sight for centuries for everyone to see, was finally recognized for what it is. This statue of the god Atlas, sentenced by Zeus to hold up the sky, holds a globe with forty-one constellations accurately placed where they would have been in the sky in 125 BC. Also, ancient coins show Hipparchus seated in front of just such a celestial globe.

A photograph of the statue was retouched, adding a strategically placed fig leaf, before it was published in a 1989 book. The *New York Times* unwittingly obtained and printed the bowdlerized photograph in its initial report of the star catalog discovery, much to the newspaper's embarrassment when the mistake became known.

Figure 4.1: Atlas Holding Up the Sky. Museo Archeologico Nazionale Napoli. Marble, seven feet tall, globe two feet in diameter, from Roman times, around 150 AD, copy of an earlier Greek work, acquired by Cardinal Alessandro Farnese, who became Pope Paul III in 1534. (Museo Archeologico Nazionale, Naples, Italy. Photo credit : Alinari/Art Resource, NY)

The Antikythera Mechanism

The Roman orator and philosopher Cicero wrote about an instrument, recently constructed, which at each revolution reproduced the motions of the Sun, Moon, and planets. Greeks, including Hipparchus (who probably died on the island of Rhodes), possessed the theoretical knowledge necessary to have built such a device, but did they also have the technological capability? Ancient written accounts of military technology on Rhodes, including a machine gun catapult with gears powering its chain drive and feeding bolts into its firing slot, were largely ignored or disbelieved by later historians.

Disbelief diminished after a sponge diver in 1900 discovered an ancient ship wrecked off the tiny Greek island of Antikythera. Along with marble and bronze statues, jewelry, pottery, and amphorae of wine were a few corroded lumps of bronze. Analysis of the pottery and amphorae suggested that they came from Rhodes and that the ship, probably sailing to Rome with its cargo, had sunk around 65 BC.

X-ray photographs of the corroded lumps later revealed about thirty separate metal plates and gear wheels. An inscription on one of the metal plates is similar to an astronomical calendar written by a person thought to have lived on Rhodes around 77 BC. This evidence pretty much ruled out the possibility that the clockwork mechanism had been dropped onto the wreck at a later date, or that it had been left behind by alien astronauts.

Cleaning away the corrosion of centuries proceeded slowly. Not until the 1950s was a more detailed analysis begun. And not until the 1970s were high-energy gamma rays used to examine interiors of the clumps of corroded bronze.

Similarities became apparent between the device and a thirteenth-century Islamic geared calendar-computer that showed cycles of the Sun and Moon on dials. Furthermore, inscriptions and gear ratios from the ancient device are similar to astronomical and calendar ratios.

In 2002, a new analysis of the device suggested that it might have used deferents and epicycles to reproduce planetary motions. In 2005, the Antikythera mechanism was examined using a prototype of a new machine designed to make three-dimensional X-rays of blades inside airplane turbine engines, and many more letters and inscriptions were found in the corroded lump of bronze. Comprehending their meaning remains a work in progress.

Greek Geometrical Astronomy

Ptolemy systematized and quantified, with rigorous geometrical demonstrations and proofs, hundreds of years of Greek geometrical astronomy. At Alexandria, Ptolemy did for astronomy what Euclid had done for geometry, and earned a reputation as the greatest astronomer of the ancient world. His mathematical systematic treatise of astronomy, *The Mathematical Syntaxis*, attracted the appellation *megiste*, Greek for *greatest*, and upon transliteration into Arabic, was preceded by *al*, Arabic for *the*—hence the *Almagest.*

In the opening pages of the *Almagest*, Ptolemy wrote that he knew he was mortal, but when he searched out the circles of the stars he stood side by side with Zeus. He would study divine and heavenly things, particularly their beautiful and well-ordered disposition, and meditate upon beautiful theorems. Echoing Plato's concern with education, Ptolemy noted that astronomy dealt with the constancy, order, symmetry, and calm associated with the divine, and thus had the potential to make its followers lovers of divine beauty and to reform their nature and spiritual state.

Ptolemy determined numerical values for his geometrical model of the solar system from many observations, and observational errors largely cancelled out. However, neither Ptolemy nor his readers understood about random errors; they would have taken every measurement as an exact result.

Therefore, Ptolemy illustrated his theory with a few observations in best agreement with it. Consequently, there are too many instances in the *Almagest* of agreement between observation and theory too good to be true—hence the recent accusation against Ptolemy of scientific fraud.

The *Almagest* was a textbook, however, not a scientific report. It taught how to convert observational data into the numerical parameters of a geometrical model. Any fudging or fabrication of observational data were little fibs allowable in the neatening up of Ptolemy's pedagogy, not lies intended to mislead his readers about crucial matters.

Ptolemy attempted to demonstrate quantitatively, using only regular and circular motions (because they are proper to the nature of divine things, which are strangers to disparities and disorders), the size and period of each planet's position relative to the Sun, and the size and period of each planet's retrograde arc. More was needed, however, to save the appearances, and Ptolemy invented a new concept: the equant point. Uniform circular motion, previously defined by cutting off equal angles in equal times at the center of the circle, would now be taken about a point not at the center of the circle. Planetary epicycles still moved with uniform angular motion, but now with respect to points other than the centers of the eccentric circles carrying them. That Ptolemy would employ such a physically implausible device as the equant point emphasizes his insistence upon accounting quantitatively for all known planetary motions.

He was not an instrumentalist inventing mathematical fictions to save the phenomena. He was a realist, and he envisioned actual physical structures in the heavens carrying around and controlling the motions of the planets. His structures, however, were not of wood, nor of metal, nor of any other earthly material; they were of some divine celestial material with no impeding nature. Thus they offered no obstruction to the passage of one part of the construction through another.

Islamic Planetary Astronomy

Alexandria had already faded to a provincial city in the Roman Empire by Ptolemy's time. Christianity triumphed in the Empire and pagan institutions were destroyed. In AD 392, the last fellow of the Museum was murdered by a mob and the Library was pillaged. Whatever remained was further damaged in the Arab conquest in 640.

Islam spread rapidly from the Arabian Peninsula, east through what is now Iran, west through North Africa by 670, and across the Mediterranean to Spain in 711. Arab scientists enjoyed a firsthand acquaintanceship with Ptolemy's *Almagest*, but were more interested in Aristotelian physical science; Ptolemy's astronomical system was little more than a convenient computing device. Nonetheless, the Islamic world did produce the most innovative addition to geometrical models of planetary motions during the Middle Ages.

In 1258, Mongol invaders under Hulagu Khan (ca. 1217–1265), a grandson of Genghis Khan (ca. 1162–1227), conquered Baghdad. The famous scholar Nasir al-din al-Tusi (1201–1274), who had been a Shiite Muslim before converting to the faith of his Sunni patrons in Baghdad, found it expedient to recant his conversion to Sunnism and to rewrite book introductions lavishly praising his former, but now vanquished, Sunni benefactors. Sunni still quarrel with Shia over the allegation that al-Tusi persuaded Hulagu to destroy the Sunni caliphate in Baghdad.

Hulagu, probably encouraged by his interest in astrology, granted al-Tusi his wish for an observatory. As expenses mounted, however, Hulagu began to doubt the utility of predicting immutable events if nothing could be done to circumvent them anyway. The observatory was constructed at Maragha, in northwest Persia (now Iran). Some of the most renowned scientists of the time, from as far as China to the east and Spain to the west, moved there, and the observatory's library was reported to have 400,000 volumes (though recent excavations have found space for far fewer volumes). Al-Tusi produced a new table of planetary positions and the Tusi couple, a combination of uniform circular motions producing net motion in a straight line.

Al-Tusi's geometrical innovation, while commanding attention in its own right, is also at the center of a historical mystery. Did Copernicus borrow this device for use in his own astronomy? Drawings by al-Tusi and Copernicus are strikingly similar, even sharing the same alphabetical letters (Arabic and Roman equivalents). Furthermore, books using the Tusi couple were circulating in Italy when Copernicus studied there at the end of the fifteenth century.

Al-Tusi had used funds from religious endowments to help finance the Maragha Observatory, suggesting that the observatory in particular and science in general were integrated into and harmonized with Muslim culture. Islam, however, was far from uniformly supportive of astronomy. Indeed, negative attitudes toward science inherent in Islamic culture may help explain why the scientific initiative shifted from the Islamic world back to the West.

Islamic traditionalists farther west than Maragha and free of the Mongol hegemony criticized al-Tusi for diverting to his observatory endowments normally devoted to institutions of charity and public assistance, such as mosques, madrases (schools), and hospitals. They also accused al-Tusi of establishing a school for heretics and teaching magic. (Harry Potter would have liked this school!) Another observatory, built at Istanbul in 1577, also came under criticism, and it was torn down shortly after its completion. The attempt to pry into secrets of nature was suspected of having brought on plague, defeats of Turkish armies, and deaths of several important persons.

Revival in the West

By AD 476, the western half of the Roman Empire had fallen to a series of invading barbarians; the eastern half survived until 1453, when Constantinople fell to the Turks. In the tenth century in the West, there was some contact with Muslim learning, and the eleventh and twelfth centuries saw an increase in translations of Arab and Greek works into Latin. Aristotelian philosophy and Christian theology fused into Scholasticism, a way of thought dominant in Western Europe between 1200 and 1500.

Scientific theories were tentative, because God could have made the world in any number of different ways all in agreement with the same observed phenomena. This nominalist thesis conceded the divine omnipotence of Christian doctrine, but at the same time established science as a respected subject and defined it in a way that, while acceding to religious authority, also removed science from under religious authority. In the new intellectual climate of nominalism, imaginative and ingenious discussions, even concerning the possible rotation of the Earth, flourished.

Hypothetical scientific theories, however, are not the stuff of revolution. The Scientific Revolution of the sixteenth and seventeenth centuries would occur only after the nominalist and instrumentalist goal of saving the appearances with mathematical fictions gave way to an impassioned quest to discover physical reality.

As early as the fourteenth century in Italy, and in universities north of the Alps in the fifteenth and sixteenth centuries, Scholasticism lost ground to Humanism and a renewed interest in Plato. Neo-Platonism, also called Neo-Pythagoreanism, included belief in the possibility and importance of discovering simple arithmetic and geometric regularities in nature, and also belief in the Sun as the source of all vital principles and forces in the universe.

A re-naissance is a re-birth, and the initial goal of Renaissance humanists was to facilitate the rebirth of Greek philosophy and values through the recovery, translation, and diffusion of lost classical works. Inconsistencies within individual ancient works and between different authors, and discrepancies between Greek scientific theories and contemporary observations initially could be attributed to defects in transmission and translation. Eventually, however, critical thought was stimulated, and what had begun as a rebirth or recovery of old knowledge transformed itself into the creation of new knowledge.

Ptolemy's *Almagest* became available to scholars in the Latin-reading world at the end of the fifteenth century. Johannes Gutenberg (ca. 1400–1468), a German metalworker and printer, introduced movable type printing around mid-century, and the German astronomer Johannes Müller (1436–1476; known as Regiomontanus) completed a new translation of the *Almagest* in 1463, although it was not printed until 1496. Nicholas Copernicus would rely heavily on the rebirth of Ptolemy's mathematical astronomy, midwived by Regiomontanus, for both its geometrical techniques and its cultural values.

Copernicus

No startling new observation in Copernicus's lifetime refuted Ptolemaic astronomy, nor did Copernicus's heliocentric system provide a better match of theory with observation. Aesthetic choices inherent within Copernicus's cultural context, including the education he received and the philosophical values he absorbed, guided him at critical junctions along his path toward revolution-making changes, if not revolution itself. His *De revolutionibus orbium coelestium* [On the Revolutions of the Celestial Spheres] of 1543 was the culmination of Greek geometrical astronomy, full of epicycles and deferents in uniform motion. Scholars have quibbled over whether Copernicus or Ptolemy employed more circles, but no one denies the presence of a plethora of circles in the geometrical schemes of both astronomers.

After his father died in 1483, Copernicus came under the guardianship of his maternal uncle Lucas Watzenrode (1447–1512). Watzenrode had attended Cracow University in Poland and the University of Bologna in Italy, where he studied church law. He then obtained a position at Frombork (Frauenburg) Cathedral (180 miles north of Warsaw), and eventually became both bishop and ruler of Warmia (Ermland), a sort of vassal state of Poland.

In 1491, Copernicus was admitted to the University of Cracow, one of the first Northern European schools to teach Renaissance Humanism. Astronomy at Cracow, however, still was taught largely in terms of Aristotelian physics. Several mathematics and astronomy books collected and annotated by Copernicus while he was at Cracow testify to his early interest in astronomy.

Watzenrode intended to appoint Copernicus to a position at Frombork Cathedral. While waiting for the next open position, Watzenrode sent Copernicus to Bologna in 1496 to study church law. The astronomy professor there, Domenico Novara, was one of the leaders in the revival of Greek geometrical astronomy and Platonic and Pythagorean philosophy.

In 1514, Pope Leo X convened a general council in Rome to study problems involving calendars, especially the date of Easter. Copernicus was invited, but he declined to attend. He replied that before the calendar problem could be resolved, the theory of the motions of the Sun and the Moon needed to be better known, and he was working on that problem.

Around this same time, Copernicus was distributing handwritten copies of his hypotheses of the heavenly motions. He began by noting that ancient astronomers had assumed that heavenly bodies moved uniformly, and they had attempted to explain apparent motions by the principle of regularity. Eudoxus had tried with concentric spheres, but failed. Many astronomers then accepted eccentrics and epicycles. The equant point, however, violated the first principle of uniformity in motion.

Copernicus's heliocentric hypothesis was the subject of a lecture given before Pope Clement VII in 1533, and in 1536, a cardinal wrote from Rome to Copernicus asking for details of his system. Attention was not all positive; a satirical play in 1531 made fun of Copernicus. More seriously, Martin Luther criticized the astronomer who wanted to prove that the Earth goes round, the fool who would turn the whole science of astronomy upside down. Holy Writ declared that it was the Sun, not the Earth, that Joshua commanded to stand still (Joshua 10:13).

Literal adherence to the Bible was the foundation of Protestant revolt against Catholic religious hegemony, and Protestants consequently rejected metaphorical and allegorical interpretations. Prior to the Counter-Reformation, in whose coils Galileo would become entangled, the Catholic Church was more liberal in its interpretation of the Bible than were Protestants, and thus more accepting of Copernican astronomy. It would be taught in some Catholic universities, and used for the new calendar promulgated by the pope in 1582.

In 1539, a young professor of mathematics, Georg Rheticus (1514–1574), showed up at Frombork Cathedral. Copernicus was almost finished with his *De revolutionibus*, and he generously shared it with Rheticus.

In 1540, Rheticus published his *Narratio prima* [First Account] of part of Copernicus's theory. Only in the heliocentric theory, Rheticus wrote, could all the circles in the universe be satisfactorily made to revolve uniformly and regularly about their own centers. Earlier planetary models were deficient in harmony, unity, symmetry, and interconnection. Ancient astronomers should have imitated musicians, who regulate and adjust the tones of strings until all together they produced the desired harmony, and established the harmony of celestial motions. All the celestial phenomena conformed to the motion of the Sun, and the harmony of the celestial motions was established and preserved under the Sun's control. The remarkable symmetry and interconnection of the motions and spheres was not unworthy of God's workmanship, nor unsuited to these divine bodies.

In the cultural climate of Renaissance Humanism, appeal to coherence and mathematical harmony must have resonated strongly among Neo-Platonists and Neo-Pythagoreans. Rheticus made the appeal explicit, writing that Copernicus, following Plato and the Pythagoreans, the greatest mathematicians of that divine age, thought that in order to determine the causes of the phenomena, circular motions must be ascribed to the spherical Earth. Divine Plato, master of wisdom, affirmed that astronomy was discovered under the guidance of God. Now Copernicus sought, with divine inspiration and the favor of God, the mutual relationship that harmonized all celestial phenomena.

Copernicus wrote in *De revolutionibus* that the ancients had not been able to discern or deduce the shape of the universe and the unchangeable symmetry of its parts, wrought by a supremely good and orderly Creator. It was as if an artist had gathered hands, feet, head, and other parts from different bodies, each excellently drawn but not matching each other, and the result was monster rather than man.

Here Copernicus was referring to the fact that several observed celestial phenomena occurred inevitably in his system but were only laboriously obtained, if at all, by intricate adjustments to the rival Ptolemaic system. Copernicus had found an admirable symmetry in the universe, a clear bond of harmony in the motions and magnitudes of the spheres, all proceeding from a single cause: the Earth's motion. Both heliocentric and geocentric theory saved the phenomena, but Copernicus's theory did so in a more aesthetic manner. Thus it was to be favored over its rival, everything else being equal.

This value judgment still guides scientists today. One of the characteristics of a satisfying theory is that observed phenomena are natural and inevitable consequences of the theory, rather than merely the product of adjustments to various parameters of the theory designed to bring it into agreement with observation.

A preface added to *De revolutionibus* without Copernicus's knowledge or approval characterized his theory as a mathematical fiction. Copernicus believed passionately, however, that the admirable symmetry and harmony in motions and magnitudes contained within his astronomical model was proof of its reality. He was confident that he had discovered the real order of the cosmos.

A Grave Matter

Copernicus was buried in a tomb under the floor of the Roman Catholic cathedral in Frombork, 180 miles north of Warsaw. In the summer of 2005, after a year of searching, Polish archaeologists located what is very likely Copernicus's grave. Police forensic experts determined that the skull belonged to a man who died at about age seventy, and a computer reconstruction of the face matches portraits of Copernicus with a broken nose and a scar above his left eye. Scientists then sought relatives of Copernicus for purposes of DNA identification. Instead, in 2008 they found in one of Copernicus's books two strands of hair, with the same genome sequence as a tooth from the skull.

For all that Copernicus accomplished, his astronomy remained the geometrical astronomy of the Greeks: clever combinations of uniform circular motions designed to save the phenomena. Switching the Earth and the Sun had little effect upon working astronomers. Either system could be used to calculate planetary positions; neither system then commanded an observational advantage. *De revolutionibus* was not a revolutionary book—it was revolution-making. How it initiated a scientific revolution is the subject of Chapter 6, The Copernican Revolution.

CONCLUSION

Continuation of cultural values from Plato to Copernicus over nearly two millennia and in different languages, countries, social and political organizations, religions, and civilizations is remarkable. Greek astronomers were neither blind nor deaf to connections between human values and science, and they realized that a sense of beauty was evoked in scientific theories. Plato regarded heaven itself and the bodies it contained as framed by the heavenly architect with the utmost beauty of which such works were susceptible. Ptolemy contemplated beautiful mathematical theories lifting him from Earth and placing him side by side with Zeus. Copernicus found admirable symmetry and harmony in the beautiful temple of the universe.

Expression of this same aesthetic sensitivity and appreciation in modern science and in the thinking of contemporary philosophers and artists is equally remarkable. The philosopher Bertrand Russell (1872–1970) noted that mathematics possesses supreme beauty, cold and austere, like that of sculpture, as only the greatest art can show.

Analogy to sculpture and to shape is not out of place in the context of Greek geometrical astronomy, infatuated as it was with the circle. According to the sculptor Henry Moore (1898–1986), there are universal shapes to which anybody whose conscious control does not shut them off responds. Is it possible that the human brain is genetically wired, and that this wiring frames the requirements for an aesthetically satisfying understanding of nature?

RECOMMENDED READING

Crowe, Michael J. *Theories of the World from Antiquity to the Copernican Revolution* (New York: Dover Publications, Inc., 1990).

Danielson, Dennis. *The First Copernicus: Georg Joachim Rheticus and the Rise of the Copernican Revolution* (New York: Walker & Company, 2006).

Dreyer, J. L. E. *History of the Planetary Systems from Thales to Kepler* (Cambridge: Cambridge University Press, 1906); 2nd ed., revised by William Stahl, with a supplementary bibliography, and retitled *A History of Astronomy from Thales to Kepler* (New York: Dover Publications, 1953).

Hetherington, Norriss S. *Planetary Motions: A Historical Perspective* (Westport, CT: Greenwood Publishing, 2006).

James, Jamie. *The Music of the Spheres: Music, Science, and the Natural Order of the Universe* (New York: Springer-Verlag, 1993).

Kuhn, Thomas. *The Copernican Revolution: Planetary Astronomy in the Development of Western Thought* (Cambridge, MA: Harvard University Press, 1957).

Toomer, G. J. *Ptolemy's Almagest* (London: Duckworth, 1984; New York: Springer-Verlag, 1984).

Saliba, George. *A History of Arabic Astronomy: Planetary Theories during the Golden Age of Islam* (New York: New York University Press, 1994).

FILM

Bronowski, Jacob, *Music of the Spheres.* 1973. 52 minutes. Color. *The Ascent of Man* series, no. 5, BBC-TV and Time-Life Films. Based on Bronowski, *The Ascent of Man* (Boston: Little, Brown and Company, 1973).

WEB SITES

Cosmic Journey: A History of Scientific Cosmology: http://www.aip.org/history/cosmology.

Nicholas Copernicus. *De revolutionibus*: http://webexhibits.org/calendars/year-text-Copernicus.html.

5

Calendars

Calendar making may not be the world's oldest profession, but the usefulness of calendars emerged early in ancient civilizations, and there is a long history of attempts to organize human lives and society in accordance with celestial time indicators.

People needed to know when winter was coming and it was time to move down from the hills to warmer valleys. They needed to know when their animals would bear young, and also when their own children would be born. They needed to know when to plant their crops and when to harvest them.

If they borrowed or lent something of value, they needed to know when to return it or when it was due. After the beginnings of a monetary economy, crop loans could be set for a year. When someone was hired for money to work for a year, both master and servant knew when the term was up without having to count the days.

Calendars also became a sign and a tool of political control. Individuals or groups who controlled calendars could claim divine power over the universe and exercise godlike power. They could manipulate time for their own benefit.

Above all, people needed to know the proper times to worship their gods, to ensure that they would have young animals, children, crops, and everything else their cultures valued. This religious need has continued throughout the history of calendars all over the world. Copernicus began his *De revolutionibus orbium coelestium* with a prefatory letter to the pope expressing the hope that his labors might contribute to the Church, especially to reform of the ecclesiastical calendar.

MARKERS OF TIME

The Sun and the Moon were the most obvious timekeepers for early civilizations. The solar, or tropical year, is the length of time for the Sun, as viewed from the Earth and envisioned in motion around the Earth, to return to the same position relative to the seasons on the Earth, a major concern for agricultural civilizations. The solar year is 365 days, 5 hours, 48 minutes, 2.9 seconds.

Even more important for early civilizations, before they evolved from nomadic into agricultural societies, was the Moon and the lunar month. The monthly phases of the Moon are more easily discerned than the slowly changing annual position of the Sun on the horizon or in the sky, and also provide a more precise measurement of time, day by day rather than season by season. A lunar month, the time between successive new moons, is 29 days, 12 hours, 44 minutes, 2.9 seconds.

The earliest calendars were very likely lunar. Nomads can observe the phases of the Moon from wherever they happen to be. The solar cycle, on the other hand, is more readily ascertained by sedentary people. With the coming of agriculture, seasons became important, and the Sun correspondingly rose in importance too.

An ideal calendar would have combined the advantages of solar and lunar timekeeping, but no number of days fits evenly into a lunar cycle, nor does any number of lunar cycles fit evenly into a solar year. Calendar development was further complicated by cultural factors, including religious considerations and political power struggles.

THE CHINESE CALENDAR

A quarter of the world's people, regardless of whether their governments have officially adopted the standard Western calendar, still affirm their cultural identity by following the ancient Chinese calendar, most notably in their celebration of the New Year. Using a traditional calendar often satisfies religious requirements, and can provide reassurance that the essence of a culture is still strong, even if the next generation is wearing blue jeans and listening to rock music.

In ancient China, control of the calendar demonstrated the power and divine connection of the emperor, and managing the calendar was a sacred task. For more than two thousand years, the emperor's bureau of astronomy made astronomical observations, prepared astrological predictions, and maintained the calendar: a union of science, pseudoscience, and government.

The ancient Chinese calendar encompassed both the solar year and the phases of the Moon, and linked its months to lunar sightings rather than to any fixed number of days. The Chinese New Year, based strictly on astronomical observations, occurs on the second new moon after the winter solstice. This solstice, according to a reform instituted in the second century BC and still in effect, must occur in the eleventh month of the year. New moon for the Chinese is the completely dark moon, the same as in the West, not the first visible crescent marking Hebrew and Islamic new moons. Chinese astronomers experimented with various patterns of adding, or intercalating, extra months to keep lunar cycles in agreement with solar years, and their schemes possibly included the nineteen-year Metonic cycle a hundred years before Meton of Athens discovered it in the fifth century BC.

The Chinese name solar years in a recurring twelve-year cycle, after their zodiacal animals (rat, ox, tiger, rabbit, dragon, snake, horse, sheep or goat, monkey, rooster or phoenix, dog, and pig). The year of the dog was AD 2006, the year of the pig 2007, the year of the rat 2008, the year of the ox 2009, and so forth.

The standard Western Gregorian calendar was introduced into China by Jesuit missionaries late in the sixteenth century, and the Republic of China formally adopted the Gregorian calendar at its founding on January 1, 1912, although warlords continued using a variety of calendars until the nationalist government consolidated its control in 1928. Chinese traditions of numbering months continued, and years were given era names beginning from 1912. In 1949, the new People's Republic of China replaced the era names with the Gregorian year counting system, eliminating a commemoration of the previous regime.

NEW WORLD CALENDARS

New World calendars did not influence development of the now standard calendar, but are greatly admired for their sophistication, and are worthy of study in their own right. They also help reveal which calendar features are determined by celestial phenomena and which are cultural artifacts.

Mayan Calendars

Mesoamerican Indians in what is now Guatemala and Mexico's Yucatan Peninsula knew the movements of the Sun, Moon, and planets, particularly Venus, against the background of the stars, and developed hieroglyphic writing and a complex calendar largely absent elsewhere in the New World. Mayan-speaking maize-farming people occupied the area as early as the fourth millennium BC. Densely inhabited villages characterize the period between 1800 BC and AD 250, followed by cities and a brilliant cultural florescence. Around 900, it all ended in collapse, one of the worst in human history, probably because war and environmental degradation left the land unable to support the large population, with severe drought the final nail in the coffin.

The Maya, obsessed with celestial cycles, discovered as many as seventeen different natural periodicities. No one else had so many different ways of keeping track of time. Religion was the major motivation, with every ritual dictated by the calendar. Mayan priests, called *He of the Sun*, computed the years, months, and days of ceremonies.

The Maya determined the length of the solar year and produced a solar calendar, with eighteen months of twenty days (the number of fingers and toes people have, not the number of days in a lunar cycle), plus an extra five days, believed unlucky. They did not add an extra day every fourth year or so, and their calendar gradually slipped out of correspondence with the seasons. They knew the length of the solar year very accurately, and knew that their calendar was slipping with respect to the seasons, but apparently they considered the calendar a sacred schedule not to be modified.

The Maya also had a ritual calendar of 260 days, consisting of a cycle of thirteen numbers and twenty names. It is still used by the Quiché of Guatemala to predict future events. This calendar was based on the average time between appearances and disappearances of Venus as a morning and an evening star, although the average periods of visibility of Venus, both as a morning star and an evening star, are 263 days, not 260. Also, in a surviving Mayan table, the times were given as 236 and 250. Why the Maya, who were outstanding astronomical observers, sanctified semi-fictitious Venus intervals is a mystery.

In 18,980 days, the 365-day and 260-day cycles return to the same celestial alignment. This 52-year cycle, called the calendar round, is still used by a few people in Mexico. Calendar round dates only place an event within the 52-year cycle, leaving ambiguity over longer periods.

The Maya had a solution for this problem: the long count of 187,200 days (5,125.36 years), the number of days since the beginning of the last great cycle. Highly elaborated calendars and texts containing long count dates are preserved on stone monuments. Supposedly cyclical creations and destructions occur at the beginning and end of each long count, and the previous cycle ended with a flood, the sky fell on the Earth, and there was no light. The current great cycle began August 11, 3114 BC and ends December 23, 2012.

Aztec Calendars

Elsewhere, the Aztecs ignored the Moon and developed an accurate solar calendar, with 18 groups of 20 days, plus 5 nothing days, making a 365-day year with a leap year every fourth year. The New Year began with the passage of the Sun directly overhead, a few days after the spring heliacal rising of the Pleiades. The Aztecs also had a second calendar, possibly adopted from the Maya, of only 260 days. As with the Maya, the two calendars created a 52-year cycle. At the end of each cycle, the Aztecs fasted and let their fires go out. Priests watched the Pleiades reach the center of the sky, assuring another 52 years. To start the new round of time, priests removed the heart from a prisoner and kindled a new fire in the resulting cavity. Throughout the country, fires were relit from this flame.

Aztec Calendar Stone

In 1749, the conquering Spaniards found this gigantic carved stone representing the Aztec calendar on the main temple at Tenochtitlan, then the Aztec capital. They tore down the temple and buried the stone under the plaza that is now Mexico City's *zocalo*, or central plaza, and the site of the city's main cathedral. The stone was rediscovered in 1790, during repairs to the cathedral.

At the center of the calendar stone is the face of the sun god, with crown, earrings, nose pendent, and necklace. The god's tongue, an obsidian knife sticking out, indicates that the god is hungry for blood and human hearts.

Around the central face are four squares, representing the four previous suns or worlds and framing the forces that ended them. The first epoch, upper right, ended when jaguars attacked and devoured the inhabitants. The other epochs were: upper left, wind (hurricanes and cyclones); lower left, rain of fire and lava (volcanic activity); and lower right, flood. The circles to the right and left of the sun god contain eagles' claws imprisoning human hearts.

In the second ring are twenty squares, for the twenty days of the Aztec month. (Clockwise, from the top: flower, rain, obsidian knife, earthquake, vulture, eagle, jaguar, reed, grass, monkey, hairless dog, water, rabbit, deer, skull, snake, lizard, house, wind, and crocodile.)

In the wedge extending down from between the monkey and the hairless dog (6 o'clock) are seven half-shut eyes, representing the seven stars of the Pleiades cluster.

Inside the square at the top of the calendar, in the outer ring, is the date Acatl 13 (AD 1479), the year the calendar was finished.

At the bottom of the outer ring are two dragon heads: the Sun and the night.

Figure 5.1: Aztec Calendar Stone. Basalt, twelve feet in diameter and weighing twenty-four tons, Museo Nacional de Antropologia, Mexico City. Courtesy of Shutterstock.

The Incan Calendar

The Incas, too, although geographically separated from the Maya and the Aztecs, used the Pleiades to mark the beginning of their new year, which began in the middle of winter (the June solstice in the southern hemisphere). The heliacal rising of the Pleiades occurred about half a lunar month before the June solstice; thus the first full moon after the heliacal rising always marked a month that includes the June solstice at the beginning of the new year. The Incas' calendar, however, was significantly different from that of the Maya and the Aztecs. It had a 328-day year of 12 lunar months of varying lengths, but averaging 27.5 days, followed by some 37 days when the Pleiades was not visible and fields were left fallow.

THE ISLAMIC CALENDAR

The Arabic predecessor to the Islamic calendar was a mixture of lunar months and solar years synchronized with the seasons by the addition of an extra month when required. The number of months had been twelve since Allah created Heaven and Earth, and it was forbidden to fight during four of these months. Inserting an extra month into the sequence could potentially transpose a forbidden month and result in eluding the timing of what Allah forbid, making lawful one year what was forbidden another year. To avoid this, Allah, through his messenger Muhammad (570–632), prohibited intercalary months.

Initially, the new Islamic calendar began with the year of the elephants. In AD 570, the same year in which Muhammad was born, an army from Ethiopia, including elephants, made an unsuccessful raid on the Kaaba, a building inside the main mosque in Mecca and Islam's holiest place. From this Year of the Elephant, later years were numbered, not only in the lunar calendar without intercalary months introduced after Allah's revelation, but also in the earlier Arabic calendar and in a lunisolar calendar introduced by Muhammad before the revelation regarding intercalary months. In 638, the caliph, Muhammad's successor as both the temporal and spiritual head of Islam, changed the numbering of years from the elephants' appearance to the Hijra, Muhammad's flight from Mecca to Medina in 622, or AH 1, or 1-H, for Anno Hegirae (in the year of the Hijra).

The Islamic lunar year, consisting of 12 lunar months of 29 or 30 days each, is about 354 days, 11 less than a solar year. Thus an Islamic holy day celebrated on a fixed day on the Islamic calendar, and also Ramadan, the month of fasting (no eating, drinking, or sexual intercourse between dawn and sunset), occur 11 days earlier every year as measured on a solar calendar and make a complete cycle around the calendar in 33.49 years.

Traditionally, the Islamic month began with the first sighting of the thin crescent Moon in the western sky after sunset by one or more trustworthy men, who then testified before a committee of religious leaders. Some Islamic countries now allow the day to be calculated in advance, but many still insist on an actual observation. Sighting the lunar crescent within one day of the new moon (the time of no visible Moon at all) is difficult, because the crescent at this time is very thin, not very bright, and easily lost sight of in the glow of the setting Sun. Usually the lunar crescent can be seen by an experienced observer on the first night after the new moon, given good viewing and sky conditions, but clouds in the sky could delay the beginning of the new month in a particular locale. Visibility is better the second day after the new moon, but even then a slender crescent very low in the west after sunset must be viewed through bright twilight in the brief time before it sets.

Saudi Arabia has several official sighting committees, and religious authorities may also declare a sighting, even if none of the committees do. Some countries begin the month at sunset on the first day that the Moon sets after the Sun, but Egypt begins a new month only if the Moon sets at least five minutes after the Sun. The farther west an observer is, the later the Moon is seen to set after the Sun; thus it is more likely that a western Islamic country will begin a new month a day earlier than an eastern Islamic country. Indeed, one year different Islamic countries ended Ramadan on each of four successive days. The date for the Hajj, the pilgrimage to Mecca, is whenever the Saudis say it is, because Mecca is in Saudi Arabia. The official Saudi calendar, however, is only for civic purposes, and in 2006 religious authorities relying on sightings of the crescent moon set the beginnings of both the Hajj and Ramadan a day earlier than called for in the civic calendar.

A tabular Islamic calendar dating back to the eighth century AD uses arithmetic rules rather than observation or astronomical calculation. Microsoft Corporation uses an algorithm based on a statistical analysis of historical data from Kuwait to convert dates between the Gregorian solar calendar and the Islamic lunar calendar.

THE PERSIAN CALENDAR

Persian astronomers, including Omar Khayyam (ca. 1048-ca. 1132), better known for his collection of poems, the *Rubaiyat* (A Book of Verses underneath the Bough/A Jug of Wine, a Loaf of Bread—and Thou), were dissatisfied with the seasonal drift of the Islamic lunar calendar, and in the eleventh century AD they devised a solar calendar. Its calculation was very difficult, however, and the Islamic lunar calendar continued in wide use in Persia through the nineteenth century.

In 1925, the beginning of the reign of Reza Shah Pahlavi (1878–1944), Persia—now Iran—officially adopted the solar calendar. Years were counted from Muhammad's flight from Mecca to Medina, as in the Islamic calendar, but the Iranians were counting solar years, not the shorter Islamic lunar years, and AH 1424 (AD 2003) was only AP 1382 (Anno Persico).

Attempting to create a monarchy-centered national mythology, in 1976 Muhammad Reza Shah Pahlavi (1919–1980) changed the first year of the Iranian solar calendar to the ascension to the throne by Cyrus the Great (ca. 590–530 BC) in 550 BC, and Iran jumped overnight from the Islamic year of 1355 to the royalist year of 2525. Reza Shah presented himself as Cyrus's legitimate heir. He also made his birthday, his son's birthday, and the date of his implementation of land reform national holidays. The murder of a prominent critic of the calendar change was widely blamed on the shah's secret police. In 1979, after the shah fled the country, Iran reverted back to counting years from Muhammad's flight to Medina.

Neighboring Afghanistan adopted the Persian calendar for their civil calendar in 1957, only to revert briefly back to the Islamic lunar calendar under the Islamic fundamentalist Taliban. Afghanistan now uses the standard Western Gregorian calendar in international relations and the Islamic lunar calendar for religious holidays.

THE WESTERN CALENDAR

Tightening ties of international travel and commerce inevitably demand more calendrical coordination between nations of diverse cultures. The standard Western Gregorian calendar, in widespread use today, is the closest thing yet to a universal calendar.

The Babylonians

The modern Western calendar originated with the Babylonians, who were among the first ancient civilizations to keep precise records of astronomical observations, which they used to develop more accurate calendars and also to predict eclipses. They began with a lunar calendar alternating months of 29 and 30 days. This was in reasonably good agreement with the actual lunar month of 29 days, 12 hours, and 44 minutes. The resulting lunar year, however, of 12 lunar months, or 354 days, was not in good agreement with the solar year of approximately 365 days.

The Babylonians devised the simple expedient of adding, or intercalating, an extra month to some solar years to make the lunar calendar come out more even with the solar year, if not every year, at least over a period of several years. They regularly added an extra month three times every eight years, and as a further adjustment, the ruler occasionally ordered insertion of an additional extra month into the calendar. Different cities added the extra month at times of their own choosing, rendering dates different from city to city.

The Babylonian month was defined as beginning with the first appearance of the crescent moon. An important duty of court astronomers was to observe and report this critical event to the king.

By around 380 BC, Babylonian astronomers knew about the Metonic cycle, having either discovered it independently or obtained it from the Greeks. Ancient Athenians had no rule for regular intercalations, and it was the task of an official to decide when to add an extra month. In 432 BC, however, the Greek astronomer Meton of Athens (fifth century BC) found that 19 solar years (6,939.602 days) are very nearly equal to 235 lunar months (6,939.688 days), with intercalations in the years 3, 6, 8, 11, 14, 17, and 19.

A century later, Callippus (ca. 370–300 BC), another Greek astronomer in Athens, improved the lunar-solar correspondence further, by taking out one day from every fourth Metonic cycle, that is, one day every seventy-six years. Callippus's calendar was used only by astronomers, who required more accuracy from their calculations than the civic calendar of Athens could provide.

Especially impressive is the Babylonian discovery of the Saros cycle (see Chapter 3, Babylonian Astronomy and Culture). The cycle was particularly good for predicting lunar and solar eclipses similar in appearance. For example, were a total eclipse followed six months later by a small partial eclipse, then after eighteen years, the length of the Saros cycle, another total eclipse would occur, to be followed six months later by another small partial eclipse. It was a calendar not of days and years, but of celestial omens, which were of great interest to Babylonians, who recorded eclipses in their horoscopes.

Jewish Calendar

The Jewish calendar, a combined solar-lunar calendar with a nineteen-year cycle, probably had its origins in the deportation of thousands of Jews to Babylon after the fall of Jerusalem in 586 BC. The modern Jewish calendar, the official calendar of the state of Israel and used worldwide for Jewish religious practice, is based on principles established by Babylonian priests and astronomers more than two millennia ago. Like others whose calendars required intercalary months, the Jews found nonastronomical reasons to insert extra months in particular places. It was convenient to insert a thirteenth month just before the beginning of spring if winter rains had not yet stopped and muddy roads would hinder Passover pilgrims, if barley was not yet ripe, or if lambs were too young and weak. Possibly a second month might have been added, were the barley still not ripe.

Egyptian Calendars

Although the origin of the modern Western calendar is attributed largely to the Babylonians, contributions from the Egyptians should not be overlooked. They began with a basic twelve-month lunar calendar of 354 days, just as the Babylonians did. But a lunar calendar could not tell in advance the most important time for Egyptian agriculture: the annual flooding of the Nile River. The heliacal rising of Sirius, the Dog Star, was noticed always to precede the flood by a few days, and the Egyptians linked it to their lunar calendar.

The Egyptian year had three seasons: flood, sowing, and harvest. There were 354 days, 11 short of a solar year. The length of the year was determined by the annual heliacal rising of Sothis (Sirius), in the fourth month of the third season (the last month of the year). If Sirius rose in the last eleven days of the twelfth month, however, then it would rise again not in the last month of the next year, but in the first month of the year after next. To avoid Sirius rising at the wrong time in their calendar, whenever Sirius rose in the last eleven days of the twelfth month, the Egyptians added a thirteenth month, making a great year of 384 days, and pulling back Sirius's next heliacal rising into the end of the following year, not the beginning of the year after.

This awkward calendar was controlled by priests, who insisted on its continued use even when a more accurate solar calendar was developed by royal astronomers. The result was two calendrical systems, one for religion and one for government.

The more accurate solar calendar consisted of three seasons of four months each, with orderly thirty-day months, plus five extra days for religious celebrations. Observers could see Sirius when it just topped an obelisk erected by astronomers, thus facilitating very accurate calculations of the time it took Sirius to return to the same position, about 365 days, the length of the year. In 237 BC, Ptolemy III further improved the accuracy of the solar calendar by adding a sixth extra day every four years, equivalent to our modern leap year. The calendar was so accurate that astronomers through Copernicus used it when constructing tables of the motions of planets.

Greek Calendars

For the Greeks, like the Egyptians and the Babylonians before them, the year consisted of twelve lunations, or new-moon to new-moon cycles, each of which lasted on average 29.5 days, and the Greeks used a lunar calendar of alternating 29 and 30 day months to keep track of everyday events.

The primary problem with a lunar calendar is that twelve lunar cycles are about eleven days less than one solar cycle. Thus without regular adjustments to the calendar, the seasons soon slip out of synchronization with the months; after only eighteen years, summer solstice would occur in December. Finding a system that reconciles the lunar year with the solar year has always been one of the great challenges of calendar making.

Most Greek cities had their own calendar, differing from others in where intercalary months were inserted, in the starting point for the year, and in the names for months. Even within a single city, calendrical confusion could reign. It was not allowed to change the day on which a feast was held, but a feast could be moved to a more convenient time by the simple expedient of strategically intercalating a day or more into the calendar.

This practice was so widespread that writers in Athens distinguished between new moon according to the goddess Selene (i.e., astronomically) and new moon according to the head magistrate of the city. In *Clouds*, the same play in which he mocked Socrates, Aristophanes (ca. 448–385 BC) had the chorus sing:

> As we were preparing to come here, we were hailed by the Moon. She said she was enraged and that you citizens of Athens treated her very shamefully, she who pays you not in words alone, but who renders you real benefits. Firstly, thanks to her, you save at least a drachma each month for lights, because as you leave home at night, you say, "Slave, buy no torches, for the moonlight is beautiful." Nevertheless you do not reckon the days correctly and your calendar is naught but confusion. Consequently the gods criticize her each time they are disappointed of their meal because the festival has not been kept in the regular order of time. When you should be sacrificing, you are putting to torture or administering justice. And often the gods are fasting while you are feasting. It is for this, that last year, when you would have made Hyperbolus king of Athens, we took his crown from him, to teach him that time must be divided according to the phases of the Moon.

For seasonal needs, the Greeks could refer to the heliacal rising or setting of bright stars or constellations. Hesiod, a Greek farmer and poet who lived around 700 BC, shared with his listeners (early poems were chanted) and readers (after the Greek alphabet was developed and poems were written down) in his *Works and Days* tips on timing farming activities by celestial appearances:

> When the Pleiades, daughters of Atlas, are rising, begin your harvest,
> and your plowing when they are going to set.
> Forty nights and days they are hidden,
> and appear again as the year moves round,
> when first you sharpen your sickle.
> . . .
> When Zeus has finished sixty wintry days after the solstice,
> then the star Arcturus leaves the holy stream of Ocean
> and first rises brilliant at dusk.
> After him the shrilly wailing daughter of Pandion, the swallow,
> appears to men when spring is just beginning.
> Before she comes, prune the vines, for it is best so.

> ...
> But when Orion and Sirius are come into mid-heaven,
> and rosy-fingered Dawn sees Arcturus,
> then cut off all the grape-clusters, Perses, and bring them home.

Labors of the Months

Figure 5.2: Labors of the Months. Carvings above the central portal of the Cathedral of Our Lady of Chartres, France, built in the twelfth century. Image copyright Mary Ann Sullivan.

The labors of the months was a pictorial convention for decorating medieval calendars, in an age when symbols were easier for an illiterate population to recognize than written names of months would have been. Each month in the solar year was identified by its zodiacal sign, and a particular activity in the annual cycle of agricultural tasks was allotted to each month. Two millennia earlier, Hesiod had similarly linked agricultural activities to celestial phenomena, particularly heliacal risings and settings of stars.

At the Cathedral of Our Lady of Chartres, built in France in the twelfth century, carvings depict a man on his knees harvesting wheat with a sickle in July, when the Sun was in the constellation Cancer, the crab. In April, the month of Aries, the ram, a man grasps a vine to prune it.

The English poet Geoffrey Chaucer also used the position of the Sun in the zodiac to indicate the month. In the prologue to his *Canterbury Tales*, he described spring as the time when the young Sun had run half his course through the constellation Aries, the ram. Also, in "The Franklin's Tale" (lines 516–521), Chaucer identified December by the presence of the Sun in the constellation Capricorn (in Chaucer's time, the Sun entered Capricorn at the winter solstice):

> The cold and frosty season of December.
> Phoebus [Apollo: the sun god] was old and colored like pale brass,
> That in hot declination [summer] colored was
> And shone like burnished gold with streamers bright;
> But now in Capricorn did he alight,
> Wherein he palely shone, I dare explain.

Show them to the sun ten days and ten nights:
then cover them over for five,
and on the sixth day draw off into vessels the gifts of joyful Dionysus.
But when the Pleiades and Hyades and strong Orion begin to set,
then remember to plow in season:
and so the completed year will fitly pass beneath the Earth.

Often, years were named after the current ruler. Alternatively, years could be counted from particular Olympiads, named by number and for the athlete who won the foot race (the *stadion*); years within the Olympiad were numbered one through four. For example, the year in which Athens restored to the Delians their island was ponderously identified by an ancient Greek historian as the year in which Astyphilos was archon at Athens, the Romans designated as consuls Lucius Quinctius and Aulus Sempronius, and the Greeks celebrated the ninetieth Olympiad, in which Hyperbios of Syracuse won the *stadion*.

Early Roman Calendars

Like most other early cultures, the Romans began with a lunar calendar. The earliest one, based on a Greek calendar, had ten lunar months, for a total of 295 days. Romans knew that the year was longer, and they suspended their calendar over a long winter gap. The Roman year began in spring, in the month of Martius, for Mars. The next three months were: Aprilis, for Aphrodite; Maius, for Maia (in Greek mythology, the eldest of the Pleiades, the seven daughters of Atlas and Pleione); and Junius, for Juno. The remaining months were just numbers: Quintilis (5), Sextilis (6), September (7), October (8), November (9), and December (10). The length of this shortened year of ten months may have been chosen because it was the approximate time for the gestation of a child, or perhaps because ten was a revered number.

The second king of Rome, who reigned from 715 to 673 BC, is credited with the addition of two more months: Januarius, for the two-faced god Janus; and Februarius, in connection with the Lupercalia, an ancient fertility feast featuring two young men racing around the city walls snapping goatskin thongs called februa (purifiers) at everyone they met.

This revised calendar still was shorter than the solar year, so the Romans tried various patterns of intercalations to keep their calendar in accord with the seasons. Some of the schemes might have succeeded had the priests in charge of announcing the intercalations not manipulated them to prolong or shorten the office terms of those they favored or opposed, or to speed up or slow down the time when debts were due. By 46 BC, neglect and greed combined had driven the calendar off its proper seasonal aspect by nearly two months.

The Julian Calendar

Julius Caesar (100–44 BC) made a serious attempt to give the Romans a calendar that accurately reflected the solar year. He may have learned about the Egyptian solar calendar from an Egyptian astronomer while having dinner with Cleopatra in Alexandria. After his return from Egypt to Rome in 47 BC, Caesar reordered the Roman calendar with the help of the Egyptian astronomer Sosigenes.

To make up for missed intercalations, the last year of the old calendar, 46 BC, was set at 445 days, with three extra months. It was called the last year of confusion, or sometimes the year of confusion. The reformed calendar had three years of 365 days followed by a leap year of 366 days. The beginning of the year was moved from March to January, to correspond with the winter solstice. The months alternated in length between 30 and 31 days, except for February, which had 29 days in a normal year and 30 in a leap year. The beginning of each month was thus strictly a function of how many days had passed since the beginning of the previous month, not the first sighting of the crescent moon.

After Caesar was assassinated, in 44 BC, the priests, not understanding the calendar and misunderstanding Caesar's instructions, began adding an extra day every third rather than every fourth year. The average year became longer and spring came earlier each year.

Caesar's nephew Augustus, who ruled Rome from 27 BC to AD 14, reformed the calendar again. He commanded that the next three leap years be omitted. He also changed the month of Quintilis to Julius, in honor of his uncle. And he changed Sextilis to Augustus, to honor himself. Thus he placed his and Caesar's names right after names of gods. July and August are better names than fifth and sixth, especially after the addition of January and February at the beginning of the year made these months seventh and eighth.

According to legend, Augustus added a day to his month, to bring it equal with Caesar's in length, at 31 days. Supposedly the extra day for August was taken from February, shortening it even more, and several other months lost or gained a day to avoid the occurrence of three short months in a row. But discovery of monuments with inscriptions showing that Sextilis already had 31 days before it was renamed August disprove this legend.

Nonetheless, and whatever the cause, instead of a simple and easy-to-remember alternation of long and short months, the modern calendar has an irregular pattern, forcing school children to memorize:

Thirty days hath September,
April, June and November.
All the rest have thirty-one,
Save February alone with twenty-eight,
But in leap year, twenty-nine.

The Romans did not officially divide their months into weeks. Instead, they had a system of markers in a month, and referred to individual days in terms of the number of days to the marker day. *Kalends*, from which comes *calendar*, was the first day of the month. *Nones* was the day the Moon reached its first quarter. And the *Ides* were the days in the middle of the month, at full moon. When the Romans switched to months with up to 31 days, they fixed the position of the Nones as either the fifth or the seventh day, and the Ides as the thirteenth or fifteenth day. Other days did not have names, but were identified by counting backwards from the Kalends, the Nones, or the Ides. Students in Latin classes are taught the mnemonic:

In March, July, October, May,
The Ides are on the fifteenth day,
The Nones the seventh; all the other months besides
Have two days less for Nones and Ides.

In Shakespeare's *Julius Caesar*, a soothsayer warned Caesar to beware the Ides of March. Shakespeare doesn't say whether his Caesar deliberately disregarded the warning or lost his life because he mistook the day in the complex Roman calendar.

The Romans did have names for the seven days of the week: the day of the Sun, the Moon, Mars, Mercury, Jupiter, Venus, and Saturn, that is, the seven celestial bodies that moved against the background of the stars. The first, second, and seventh of these days are now Sunday, Monday, and Saturday. Current English names for the other days of the week reflect a migration from the Roman system to Norse mythology: Tiw, Woden, Thor, and Freya; or Tuesday, Wednesday, Thursday, and Friday. Nearby cultures placed more emphasis on a seven-day cycle than did the Romans; worshipers of Saturn set aside Saturn's Day, and the Jews took the seventh day as holy.

Constantine

Although esteemed as the first Christian emperor and venerated as a Christian hero, Constantine (ca. 274–337) might be more accurately acclaimed as a man who knew how to use religion and the calendar to bolster his own power. He defeated a rival in battle after purportedly seeing in the sky a flaming cross with the words (in Greek) *In This Sign Conquer* and after Christ appeared in a dream to give Constantine advice about the insignia for his troops. Also, Constantine made the seven-day week official and declared the day of the Sun a holy day, on which courts would be closed and only farmers should work. Thus the anti-Semitic emperor made sure that the holy day was different from the Jewish Sabbath, and at the same time offered a sop to powerful Sun worshipers, including the Mithraists, who then outnumbered Christians.

Constantine worked to solidify his power by insuring that the bishops presented a unified front among themselves and with the state. A unified church, like a unified government, would be stronger. To this purpose, Constantine in 325 convoked the first ecumenical conference of bishops, which produced the first uniform Christian doctrine, the Nicene Creed, so named after the meeting place, in Nicaea (now Iznik, Turkey).

One of the major successes of the meeting was the synchronization of the timing of the Church's most important celebration, Easter. The Last Supper was clearly the Passover meal, and the bishops might have tied Easter to Passover. Constantine, however, declared that they should have nothing in common with the Jews. He wanted an astronomical method that the Church could compute itself. The bishops decided that Easter would be the first Sunday after the fourteenth day of the Pascal Moon, the new moon whose fourteenth day followed the vernal or spring equinox.

It is relatively easy to watch for the equinox and the next full moon. But to prepare for celebrations, the Catholic Church needed to know the date of Easter years in advance. And to calculate the date of Easter, the Church needed to know average times between successive full moons and between successive vernal equinoxes. Difficulties predicting the lengths of lunar months and the timing of the equinox would be central in the continuing history of the Western calendar.

Dionysius Exiguus and Anno Domini

Constantine's efforts to unify the empire and the church did not last. After the empire split into eastern and western halves, the Latin Western Church used Julius Caesar's date of March 25 for the equinox and an eighty-four-year cycle of lunar months. Meanwhile, the Greek Eastern Church calculated Easter using March 21 as the official date of the spring equinox and a nineteen-year Metonic cycle.

Rome tried to reform its system in the sixth century, when Pope John I asked Dionysius Exiguus (ca. 470–544), a Russian monk, to produce a new table of Easter dates. Dionysius claimed that the Holy Spirit was the source of his new table, but modern historians believe he used mathematics and the nineteen-year Metonic cycle. An existing table covered the nineteen-year period denoted 228–247, with years counted from the beginning of the reign of the Roman emperor Diocletian. Dionysius continued the table for another nineteen-year period, which he designated Anni Domini Nostri Jesu Christi 532–550.

The calendar year abbreviation AD originally stood for Anno Diocletiani, the year of Diocletian, and Egypt's Coptic Christians still count calendar years in this manner. The Coptic calendar is particularly convenient for commemorating martyrs created by Diocletian, the last Roman emperor to order major persecutions of Christians. According to the Gregorian calendar, Diocletian's reign began in 284; thus the Coptic calendar is 284 years behind the Gregorian calendar.

Dionysius reset the beginning of the calendar to the birth of Jesus Christ, whom he decided had been born 531 years earlier (which made the then current year AD 531, for Anno Domini, in the year of our lord). The book of Matthew, however, states that Jesus was born during the reign of Herod the Great, and Herod died in 4 BC.

There was no year zero because the numbering system lacked a zero; not until the eleventh century would zero be used regularly in Europe. Lack of a year zero also meant that the millennium would end not with the year 999, but in the following year. Thus in 1999, when many people referred to the coming year as the New Millennium, others gleefully corrected them, insisting that the New Millennium would not arrive until 2001.

Recent efforts to switch from BC (Before Christ) and AD (Anno Domini) to BCE (Before Common Era) and CE (Common Era) represent a compromise, between an after-the-fact disinclination to tie the time-keeping system to a specific religion and reluctance to change a now long-standing and widely used numbering system for calendar years.

Dionysius's work also strengthened the accepted date for Christmas. Roman Christians celebrated the birth of Jesus on December 25, conveniently the same day as the major Roman holiday celebrating the birth of the god Mithras. Christians in Antioch, however, celebrated Jesus's birth on January 6. And many didn't celebrate it at all. Dionysius argued that God had created the Earth on March 25; that Jesus, a perfect being, must have been conceived on March 25; and thus must have been born on December 25. There is no historical evidence for December 25, but the date has become firmly embedded in countries with large Christian populations and large retail sales.

Aborted Attempts toward Calendar Reform

Roger Bacon (ca. 1214–1292), an English philosopher and Franciscan friar, was a leading scientist at Oxford before he disagreed with the new head of his order's scientific branch and was banished to a monastery in France. There, Cardinal Guy le Gros de Foulques became interested in Bacon's ideas, and asked him to write a comprehensive treatise. But this was against the rules of the Franciscan order. The cardinal, upon becoming Pope Clement IV, told Bacon to ignore the prohibition and secretly write his book, on how the philosophy of Aristotle and the new science could be incorporated into a new theology.

Included in Bacon's *Opus majus* [Major Work], was his calculation that the calendar then in use, established by Julius Caesar, averaged eleven minutes a year more than the actual solar year, and thus had fallen out of correspondence with the seasons by nine days since Caesar's time. Bacon

Figure 5.3: *The Cycle of Life*, woodcut, 1504. The outer ring contains the signs of the zodiac and the inner ring the corresponding seasonal labors (the same as the Labors of the Months in figure 5:2). In the center, a woman holding flowers against her womb represents fertility. The man in the barren landscape is comforted by fire. Rare Book and Special Collections Division, Library of Congress.

also noted that the equinoxes were not occurring at the same time every year, but ascending through the calendar. Unfortunately, Clement died before he could read Bacon's book. In 1345, Pope Clement VI developed a plan to reform the calendar, but then he died of bubonic plague, again delaying calendar reform.

Zodiac Man

In much the same manner that each month in medieval calendars was indicated by its zodiacal sign and allotted a particular agricultural task, each constellation was also thought to govern a particular region of the human body, which should not receive medical treatment during the corresponding time period. Astrology linked the body's microcosm to the macrocosm of the planets and stars, whose movements were thought to influence movements below the orbit of the Moon. Ancient Greek astrological thought was preserved by Islamic writers and passed back to the West between the twelfth and fifteenth centuries AD.

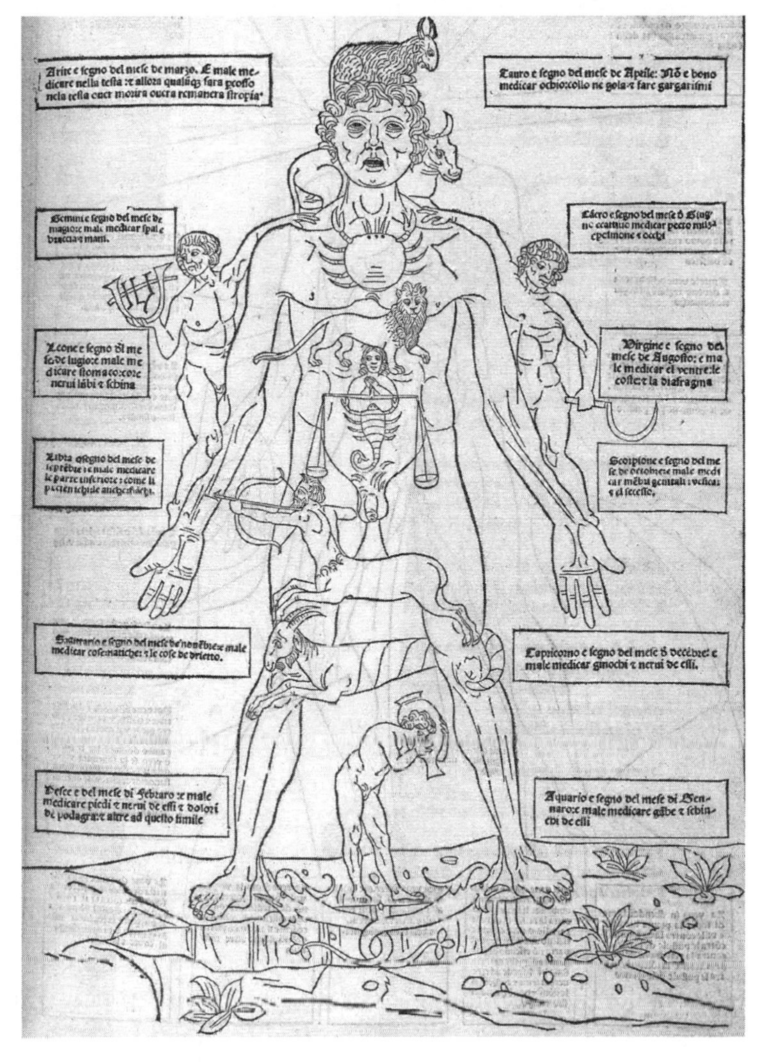

Figure 5.4: *Zodiac Man*, woodcut, 1493. From *Fasiculo de medicina*, medical writings collected by Johannes de Ketham, a German physician working in Venice, Italy. National Library of Medicine.

Aries (ram) governs the head, eyes, adrenals, and blood pressure, and anyone who treated the head in March or April would cause a concussion or death. Taurus (bull) governs the neck, throat, shoulders, and ears; Gemini (twins) the lungs, nerves, arms, and fingers; Cancer (crab) the chest wall, breasts, and some body fluids; Leo (lion) the heart, spine, upper back, and spleen; Virgo (maiden) the abdomen, intestines, gallbladder, pancreas, and liver; Libra (beam balance) the lower back, hips, kidneys, and endocrines; Scorpio (scorpion) the reproductive organs, pelvis, urinary bladder, and rectum; Sagittarius (archer) the thighs and legs; Capricorn (goat) the knees, bones, and skin; Aquarius (water bearer) ankles and blood vessels; and Pisces (fish) feet and some body fluids.

Chaucer wrote in the prologue to his *Canterbury Tales*:

With us there was a doctor of physic [medicine];
In all this world was none like him to pick
For talk of medicine and surgery;
For he was grounded [instructed] in astronomy.
He often kept a patient from the pall [death]
By horoscopes and magic natural.
Well could he tell the fortune ascendent [calculate the planetary position]
Within the houses [the twelve divisions of the zodiac] for his sick patient.

A Sheepherder's Life in Astronomical Context

The English poet Edmund Spenser (1522–1599) in his 1579 *Shepheardes Calender* characterized the life of a shepherd in terms of the four seasons: springtime for youth; summer for love; fall with ripe fruits; and winter showing the shepherd's latter years as chill and frosty. For passionate summer, Spenser wrote:

> And Summer season sped him to display
> (For love then in the Lions house did dwell)
> The raging fire, that kindled at his ray.
> A comet stirred up that unkindly heat,
> That reigned (as men said) in Venus' seat.

The planet Venus was often linked with the goddess of love. The lion's house was the constellation Leo, indicating the summer month of August. Further on, a line drawing depicted the shepherd under a goat-shaped cloud in the sky, referring to the constellation Capricorn, the goat, and indicating that it was the month of December.

The Gregorian Calendar

The calendar dictated by Julius Caesar in the first century BC was still in use at the beginning of the sixteenth century AD. Its year was 365 days long, with an extra day added to the month of February every fourth year, without exception. The actual length of the year, however, is about eleven minutes less than Caesar decreed, which works out to an error of one whole day every 128 years, as Bacon had realized.

This would not have mattered much to the Church, had its liturgical calendar used the Julian solar calendar exclusively. But in AD 325, under Constantine's guidance, Easter had been linked to the Moon rather than to the Sun. Translating a fixed date from a lunar calendar to a variable date in a solar calendar containing an error of several days over several centuries is not easy, and, to its considerable embarrassment, the Church was unable to predetermine very far in advance the Julian date of Easter.

For this reason, Pope Leo X convened in Rome in 1514 a general council to study problems involving calendars, especially the date of Easter, and invited Copernicus to participate. He declined, replying that before the calendar problem could be resolved, the lengths of the years and the months and the motions of the Sun and the Moon needed to be better known, and he was working on that.

In 1543 his work was complete, and in a prefatory letter in his book *De revolutionibus* addressed to Pope Paul III, Copernicus expressed his hope that his labors contributed somewhat even to the Commonwealth of the Church. Two years later, Paul initiated the Council of Trent, a commission of cardinals to plan institutional reform. In 1563, under Pope Pius IV, the council declared it necessary to reestablish the date of the vernal equinox at March 21 and to keep the calendar from drifting in the future.

Catholics viewed the reform movement as a Catholic reformation, but Protestants focused on the aspect of reaction to Protestant reformers and labeled the movement the Counter-Reformation. Accordingly, Protestant states were likely to oppose any proposed calendar changes, and the new calendar introduced in 1582 by Pope Gregory XIII would be adopted more readily in Catholic countries than in Protestant ones.

Gregory reestablished the date of the vernal equinox at March 21 by dropping ten days from the calendar. October 4, 1582 was followed by October 15, 1582. Most of the drift of the date of the vernal equinox was eliminated with the elimination of the century leap years that were not also divisible by 400. This removed three days every 400 years and eliminated all but about 0.0003 day of error per year.

The Sun in the Church

To prepare for celebrations, the Catholic Church needed to know the date of Easter years in advance. And to calculate the date of Easter, the Church needed to know the average times between successive full moons and between successive vernal equinoxes. The essential measurement is how long it takes for the Sun's noon image to return to the same place on a north-south line (the meridian). Ideally, the line is laid out in a dark building with a small hole high in a wall or in the roof to admit a narrow ray of sunlight. The longer the north-south dimension of the building, the longer the meridian line and the movement of the Sun's noon image, and consequently the more accurate the measurement.

All this explains why, late in the sixteenth century, a tiny hole was made in the great circular window seventy feet above the floor of Santa Maria Novella, the first of the great basilicas in Florence. The meridian line ran down the length of the nave of the cathedral, and the Sun's image differed in location along the meridian line by more than a hundred and fifty feet from winter to summer solstice. Essentially, this church was now a huge solar observatory. The same was attempted with Bologna's great basilica of San Petronio. There, piers supporting the nave blocked the first attempt, but eventually this church, too, would yield precise astronomical measurements. Though used as solar observatories, these churches had not been built with that purpose in mind. Nor did their priests and congregants worship the Sun.

This disappearance of days left renters by the month paying thirty-one days of rent for twenty-one days of use, and not everyone was happy. Protestant theologians also objected, calling Gregory's calendar a Trojan horse designed to trick real Christians into celebrating Easter at the wrong time. Greek Orthodox churches were hostile, too, to pronouncements from Rome, and Gregory's calendar reform was condemned as contrary to tradition, to the Scriptures, to councils, and to the wishes of the Church founders; furthermore, the reform was simply a vanity of the pope. Many Orthodox churches continue to celebrate Easter according to the Julian calendar.

In England, Queen Elizabeth's astrologer, John Dee (1527–1609), also a mathematician, but more respected in his lifetime as the former than the latter, proposed a similar calendar reform, removing eleven days to bring the calendar into line with the astronomical year. Elizabeth's court generally approved, but the Archbishop of Canterbury was opposed, because he distrusted Rome. He may also have seized the opportunity to thwart a queen not always in agreement with his plans for reorganizing the English church.

Most of Italy, Spain, and Portugal immediately adopted the Gregorian calendar, and other Catholic countries soon followed. Not until the beginning of the eighteenth century, however, did Denmark and Protestant regions in Germany begin using the Gregorian calendar.

In Sweden, a decision was made to migrate the calendar slowly from Julian to Gregorian, by leaving out each leap day correction beginning in 1700 until 1740, and to continue to celebrate Easter on its Julian date until 1740, after which it would be celebrated on an improved, astronomical date. But Swedish almanac makers mistakenly inserted leap days for 1704 and 1708, and to end the confusion, it was decided in 1712 to revert back to the Julian calendar, by adding an extra leap day, February 30 (the only genuine February 30 in history), for the missed leap day of 1700. Sweden remained on the Julian calendar until 1753. The improved dating of Easter was introduced in 1740, as scheduled; consequently, Sweden celebrated Easter a week later than the rest of Europe until 1844, when the country formally adopted the Gregorian Easter reckoning.

Britain and its colonies converted in 1752, and the day following September 2 became September 14. Dissatisfaction with the change surfaced in the next election, and William Hogarth's 1754 painting and print *An Election Entertainment* includes a banner with the campaign slogan "Give Us Our 11 Days." Special legislation in Britain ensured that monthly and yearly payments would not become due earlier than they would have been using the Julian calendar, with the result that the beginning of the next tax year was changed from March 25 to April 5. Another Julian leap day was dropped in 1800, and the tax day accordingly changed from April 5 to April 6. But the tax day was not changed again in 1900. Consistency with the rest of the world was the decisive argument in conversion from the Julian to the Gregorian calendar in Britain, whose international trade had become the economic lifeblood of many of its citizens.

In 1793, France adopted a new calendar designed to exemplify the rationality of its new republic. The counting of years started from the proclamation of the republic as year one; there was no year zero. There were 12 months of 30 days each, with 5 more days for festivals and vacations, and a sixth festival day on leap years. Each month had 3 periods of 10 days each, with the tenth day a day of rest. The plan also called for days to be divided into 10 hours of 100 minutes each, but available clocks and watches were not adequate to this decimal task. The months were renamed Vendemiaire (vintage), Brumaire (mist), Frimaire (frost), Nivose (snow), Pluviose (rain),

Different Times in Different Places

Miguel Cervantes, the author of *Don Quixote de la Mancha*, died on April 23, 1616, in Spain, under the Gregorian calendar, and William Shakespeare died in England ten days later, on April 23, 1616, under the Julian calendar. William III of Orange sailed from Holland on November 11, 1688, for Britain, to become that nation's new king, and arrived on November 5, 1688, after four days at sea. Easter Island was discovered by a Dutch sailor on April 5, 1722, the Gregorian calendar day for Easter; had a British sailor discovered the island on the same day, March 25 in the Julian calendar, he might have named it Annunciation Island or Lady Island, because March 25, was Lady Day in Britain, the day of the Annunciation by the archangel Gabriel to Mary that she would bear Christ. George Washington celebrated his birthday on February 11 before conversion from the Julian to the Gregorian calendar in Britain and its colonies, and thereafter on February 22.

Ventose (wind), Germinal (seedtime), Floreal (blossom), Prairial (meadow), Messidor (harvest), Thermidor (heat), and Fructidor (fruits). Difficulties converting dates between the revolutionary and the Gregorian calendars and communicating with other countries persuaded Napoleon to return France to the Gregorian calendar at the beginning of 1806.

Japan came to appreciate the power of Western weapons and technology, and under the emperor Meiji, who reigned from 1852 to 1912, launched an all-out effort to pattern its entire economy and society after the West. As part of its Westernization, Japan converted from the Chinese lunar calendar to the Gregorian solar calendar in 1873, but with a difference in how the years are counted. Historically, the Japanese had begun numbering years anew every time a significant event was deemed to have occurred. Since the Meiji reform, however, a new era has started only on the ascension of a new emperor. While the Gregorian calendar in the West is now into years in the two thousands, the longest era in the Japanese calendar is the sixty-four-year Showa era. The era of Akihito, the current emperor, is known as Heisei, meaning peace everywhere.

In 1829, the Russian Academy of Science petitioned the czar to accept the Gregorian calendar in place of the Julian. Any possible advantages of the proposed reform, however, appeared to the czar relatively small in comparison with potential inconveniences and difficulties, very likely including upheavals and bewilderment of mind and conscience among his people. Using different calendars complicated coordination of military initiatives between Russia and her allies during World War I, although the Russian Department of Foreign Affairs, the commercial and naval fleets, and scientists, including astronomers, had already been using the Gregorian calendar in their relations with foreign countries.

In 1918, Lenin and the new Bolshevik government replaced the Julian calendar in Russia with the Gregorian, necessitating cancellation of thirteen days. The Russian Orthodox Church, partly in protest against the Bolsheviks and partly from reluctance to change, stuck with the Julian calendar.

Consequently, religious Russians celebrating Christmas on December 25 (Julian date) are seen by others as celebrating Christmas on January 7 (Gregorian date).

In 1923, Russia switched to a new calendar, replacing all religious feasts and holy days with five national public holidays associated with the 1917 Bolshevik Revolution. In an effort to increase production, the Communists first tried a system of five-day weeks and six-week months, with every day except the five holidays a workday. Workers had staggered days off, so production never stopped. This arrangement was particularly hard on family life, and in 1931, the calendar was changed to a six-day week with the sixth day as a rest day. This calendar lasted until 1940, when Russia resumed use of the Gregorian calendar.

The date of the beginning of the year also has varied in Russia. Until the end of the fifteenth century, when Ivan III relocated the Russian capital to Moscow, years were counted from the creation of the world (5509 BC) and the Russian year began on March 1, as had the old Roman year, and as it did in Venice until the conquest of that republic by Napoleon in 1797. Ivan's new government in Moscow changed the beginning of the year to September 1, and in 1699, Peter the Great, who instituted many reforms designed to modernize Russia and make it a European nation, changed the beginning of the year to January 1.

CONCLUSION

To a large extent, periodicities in the natural world, particularly the phases of the Moon and the seasons of the solar year, inevitably have driven civilizations toward a virtually predestined calendar, not withstanding many cultural obstacles and distractions along the way. Astronomers and others involved in calendar development and reform could not foresee the future, of course, but nonetheless were guided forward by celestial phenomena adopted for time keeping. Indeed, calendar schemes are so similar across ancient cultures that some people have even taken this as evidence that alien astronauts visited various ancient civilizations and left behind calendars.

In retrospect, lunar calendars were a historical blind alley. Now, when used at all, lunar calendars are primarily religious relics, although nonetheless important to celebrants of religious rituals. The Moon's most lasting effect on the modern calendar seems likely to be the dozen lunar months, approximately, in a solar year.

There is, of course, no need to tie months tightly to the Moon. Thirteen months of twenty-eight days each, seven days in each week, plus an extra day at the end of the year, would satisfy the Vatican's call for a perpetual civil calendar keeping Sunday the first day of every month and every week. Proponents of the Thirteen Moon Natural Time Peace Calendar argue that it would also maintain a closer link than other calendar systems to natural human biological cycles. Nor would it lack religious advantages, including a connection with the Mayan prophecy of the end of the world in 2012 and encouragement of a time of real harmony and peace on Earth through calendar reform.

How the days of the month are apportioned among weeks or some other unit, if any, is a historical and cultural artifact, not astronomical, although phases of the Moon can be used as an indication of how far the monthly lunar cycle has progressed in days. Cycles of several days of work ending with a market day, a day of rest, or a day of worshipping a deity characterize many societies and their calendars. A nursery rhyme fills the days of the week with different tasks: this is the way we wash our clothes so early Monday morning, iron our clothes so early Tuesday morning, mend our clothes Wednesday, scrub the floors Thursday, sweep the house Friday, bake bread Saturday, and go to church Sunday. At the end of the Incas' week, the king changed wives.

Divisions of the day into hours, minutes, and so forth might be better done decimally, as the French rationalists in their revolutionary fervor attempted, albeit without lasting success. Undoubtedly, this is the logical way Spock of the television series *Star Trek* marks the passing of a day on Vulcan.

The solar year has proven of more lasting significance, although the declining percentage of the population engaged in agricultural activities raises questions for the future. Maybe increases in seasonal recreational activities will continue demand for a solar calendar.

Ultimately, consistency is probably more valuable than any other consideration, as human interrelationships are increasingly internationalized. This

Leap Seconds

The essential problem of calendars for people on Earth has not really been solved; it has just become more refined. Humankind has learned to use leap years and leap days to keep the human year in alignment with the Sun, but now there are also leap seconds to keep the day aligned. The day as measured by an atomic clock does not always match the day as measured from noon to noon; the physical day is not always the same length, because the pull of the Moon is constantly slowing the Earth's rotation. Leap seconds are used to slow clocks to match more closely the actual time of the Earth's rotation. However, as with historic manipulation of the calendar, there are practical and social obstacles creating stumbling blocks in the way of a system that would make clocks match the sky.

It is difficult to design computer software to handle the extra seconds, because—just as the medieval Church was unable to predict the spring equinox with sufficient accuracy to set Easter very far in advance—modern astronomers are unable to predict when a leap second will be needed to synchronize human clocks with the Earth's rotation. Experts dealing with air traffic control, electronic fund transfers, and satellite communications have argued that leap seconds do more harm than good and that a difference in timekeeping systems as small as a second could cause an airplane crash.

is probably why Britain adopted, however begrudgingly and belatedly, the Gregorian calendar, and why countries still using other religious calendars simultaneously use a Gregorian calendar in civil and commercial dealings with the rest of the world. *Star Trek* fans may fantasize that the star date system of time, introduced shortly after the founding of the United Federation of Planets in 2161 AD, standardizes time measurement among various cultures, worlds, and races, enabling planets and outposts light years apart and starships traveling at warp speeds (greater than the speed of light) to keep track of a unified time base despite relativistic effects, universal expansion, and the effects of gravity on time and space. The zero point of the system was arbitrarily set at the date of incorporation of the Federation.

RECOMMENDED READING

Aveni, Anthony. *Empires of Time: Calendars, Clocks, & Cultures* (New York: Basic Books, 1989).

Burland, C. A., and Werner Forman. *Feathered Serpent and Smoking Mirrors* (New York: G. P. Putnam's Sons, 1975).

Coe, Michael D. *The Maya*, 7th ed., fully revised and expanded (New York: Thames & Hudson, 2005).

Duncan, David Ewing. *Calendar: Humanity's Epic Struggle to Determine a True and Accurate Year* (New York: Avon Books, Inc., 1988).

Evans, James. *The History and Practice of Ancient Astronomy* (New York: Oxford University Press, 1998).

Milbrath, Susan. *Star Gods of the Maya: Astronomy in Art, Folklore, and Calendars* (Austin: University of Texas Press, 1999).

WEB SITES

Aristophanes. *Clouds*: http://classics.mit.edu/Aristophanes/clouds.html.

Calendar Reform: http://personal.ecu.edu/mccartyr/calendar-reform.html#BB.

Calendars through the Ages: http://webexhibits.org/calendars/index.html.

Chaucer, Geoffrey. *The Canterbury Tales*: http://www.canterburytales.org/canterbury_tales.html.

Hesiod. *Works and Days*, translated by H. G. Evelyn-White: http://www.theoi.com/Text/HesiodWorksDays.html.

Spenser, Edmund. *The Shepeardes Calender*: http://uoregon.edu/~rbear/shepeard.html.

6

The Copernican Revolution

Revolutions in science often surpass the limited changes envisioned by their initiators, much as political and social revolutions do. The Copernican revolution began with a simple exchange of the Earth and the Sun in a geometrical model still composed of uniform circular motions. The astounding implications of this change appeared only later.

The prevailing physics, in which objects fell to the Earth because of their natural tendency to move to the center of the universe, where the Earth was, did not work in a Copernican universe. Thus Aristotelian physics would be one of the first casualties of the revolution.

Furthermore, other worlds, their existence previously forbidden because their earthy nature would have caused them to move to their natural place at the center of the universe, now could be scattered throughout a Copernican universe. The Earth was one of many planets similar in physical composition, and possibly with similar inhabitants. No longer could the universe be thought of as specifically created for humankind, nor humans as unique and special. Many intelligent species might be scattered throughout the universe.

The revolution would also see the demise of the finite, closed, and hierarchically ordered universe of ancient and medieval belief. It gave way to an indefinite or even infinite universe consisting of components and laws, but lacking value concepts, including perfection, harmony, meaning, and purpose. While scientists were changing understanding of the physical universe, other thinkers radically altered political, sociological, and religious views of humankind's place and role in the new universe.

DISTRIBUTION OF STARS IN SPACE

Attribution of the apparent motion of the stars to rotation of the Earth rather than to rotation of the outer sphere of the stars rendered that sphere obsolete. A logical consequence, albeit unrealized by Copernicus, was a newly possible distribution of stars throughout a possibly infinite space.

The change in thinking is nicely illustrated in the work of Leonard and Thomas Digges, father and son. Leonard (ca. 1520–1559), an English mathematician and surveyor, took part in an unsuccessful rebellion in 1554 against the new Catholic Queen Mary and was sentenced to death. He ended up forfeiting not his life but his estate, which had supplied him with ample means and leisure. His almanac, *A Prognostication Everlasting,* first published in 1553, was a welcome source of now desperately needed income. He published a second edition in 1555, and then a third in 1556. It contained the usual calendar and weather facts or fables, and also astronomical information, including a standard medieval diagram of the universe with the Earth in the center.

After Leonard's death, his son Thomas (1546-1595) became a ward of Queen Elizabeth's astrologer John Dee, and under Dee's tutelage became an astronomer. In 1576 Thomas reprinted his father's *Prognostication Everlasting* without change in the text from the final, 1556 edition, but with a new appendix. Thomas described the great orb of the heavens as "garnished with lights innumerable and reaching up in *Sphaericall altitude* without end." A diagram depicted stars scattered at various distances beyond the former boundary of the sphere of the stars.

Thomas Digges' diagram also illustrates continued obliviousness to another logical consequence of the Copernican model. Digges accorded to the Sun a unique place at the center of the universe, even though the concept of a center loses all meaning in an infinite universe.

Approximately two thousand different editions of almanacs were published in England during the following century, and an estimated three to four million copies total. With the sole exception of the Bible, an almanac was the book most commonly found in English homes. Only rarely, however, were new astronomical discoveries mentioned in almanacs, because few readers cared whether the Sun or the Earth revolved around the other; they wanted prognostications of the future, as revealed in the changing positions of the Sun, Moon, and planets as measured against the constellations.

GALILEO

For impact on the great debate over the Copernican world view, nothing else came near to equaling the telescopic discoveries of Galileo Galilei (1564–1642). They did not prove the Copernican theory, but in destroying

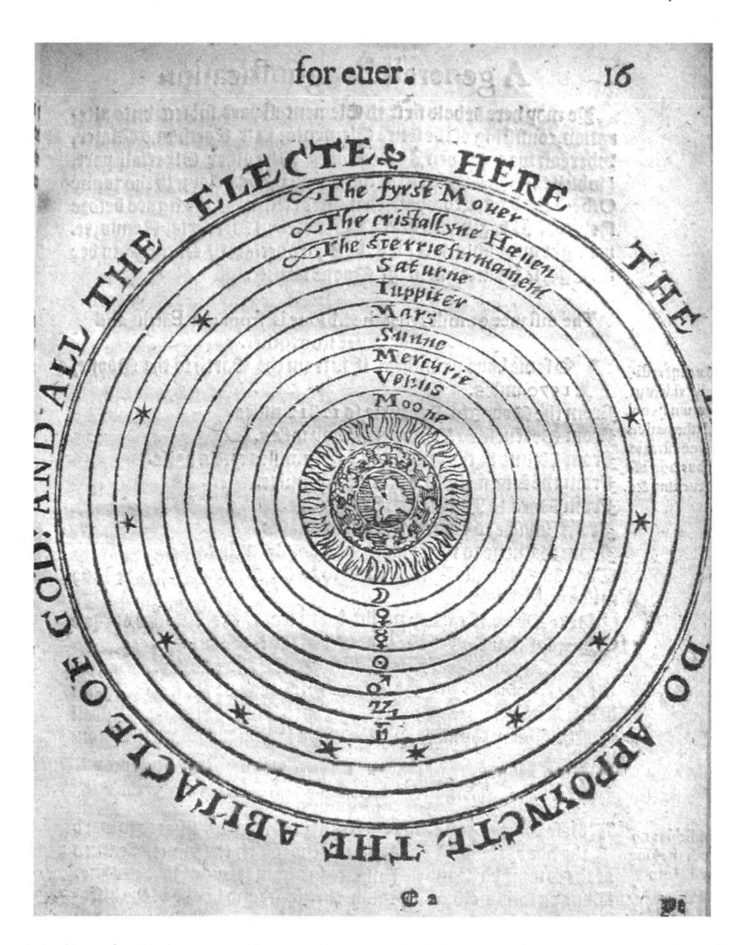

Figure 6.1: Medieval Universe. From Leonard Digges, *A Prognostication Everlasting*, third edition, 1556. The sub-lunar region consists of spheres of earth, water, air, and fire. Beyond are the spheres of the Moon, Sun, planets, and stars. Image courtesy of History of Science Collections, University of Oklahoma Libraries; Copyright the Board of Regents of the University of Oklahoma.

its alternative, the ancient Aristotelian cosmos, Galileo's observations implicitly supported the Copernican universe.

Galileo was born in the same year as Shakespeare, and in the year of Michelangelo's death. Galileo's father wanted him to study medicine, and he did so at the University of Pisa. But he preferred mathematics, and he studied it with private tutors in Pisa and at home, in Florence. Soon Galileo was giving private lessons. Next he was appointed to the mathematics chair at the University of Pisa, when he was only twenty-six years old. Obstinate and argumentative, he was unpopular with other professors. In 1592, he left Tuscany to take up the chair of mathematics at the University of Padua, in the Republic of Venice, where he stayed for eighteen years.

In 1609, rumor reached Galileo from Holland of a device using pieces of curved glass to make distant objects appear near. He quickly constructed his own telescope and turned it to the night sky. Large dark spots on the

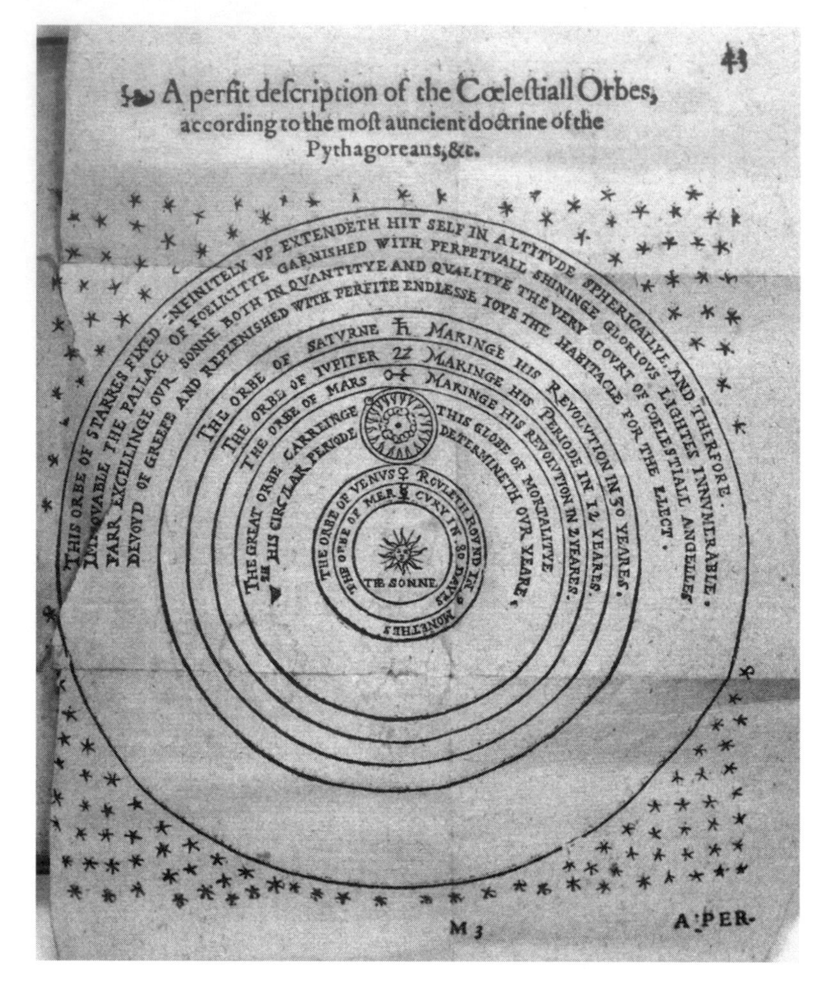

Figure 6.2: Copernican Universe. From Thomas Digges, *A perfit description of the Caelestial Orbes, according to the most ancient doctrine of the Pythagoreans, & c.*, 1576. The universe is centered on the Sun, and the stars are scattered throughout a vast open space. Thomas added this diagram as an appendix to his father Leonard Digges' earlier almanac: *A Prognostication Everlasting*. Image courtesy of History of Science Collections, University of Oklahoma Libraries; Copyright the Board of Regents of the University of Oklahoma.

Moon had been observed before Galileo, but his more precise and detailed telescopic observations emphatically demanded the revolutionary conclusion that the Moon was not a smooth sphere, as Aristotelian physics maintained, but that its surface was uneven and rough, like the Earth's.

Next, Galileo turned his telescope on Jupiter, and found four moons circling the planet. This observation suggested that the Earth is a planet, just like other planets with moons circling them. Here was a fine and elegant argument, Galileo wrote, for quieting the doubts of those who were mightily disturbed to have the Moon alone revolve around the Earth and

Early Drawings of the Moon

The boundary or terminator dividing the dark and light parts of the Moon is not a uniform oval line, as it would be for the smooth sphere demanded by Aristotelian physics. It is uneven and wavy because the surface of the Moon is uneven, rough, and full of cavities and prominences. The Moon's surface is like the Earth's, with mountains and valleys.

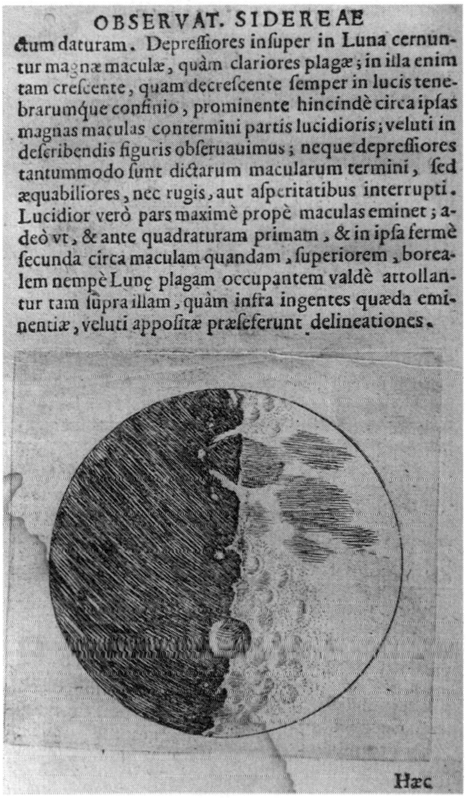

OBSERVAT. SIDEREAE

ctum daturam. Depreſſiores inſuper in Luna cernuntur magnæ maculæ, quàm clariores plagæ; in illa enim tam creſcente, quam decreſcente ſemper in lucis tenebrarumque confinio, prominente hincindè circa ipſas magnas maculas contermini partis lucidioris; veluti in deſcribendis figuris obſeruauimus; neque depreſſiores tantummodo ſunt dictarum macularum termini, ſed æquabiliores, nec rugis, aut aſperitatibus interrupti. Lucidior verò pars maximè propè maculas eminet; adeò vt, & ante quadraturam primam, & in ipſa fermè ſecunda circa maculam quandam, ſuperiorem, borealem nempè Lunę plagam occupantem valdè attollantur tam ſupra illam, quàm infra ingentes quæda eminentiæ, veluti appoſitæ præſeferunt delineationes.

Hæc

Figure 6.3: Drawing of the Moon. From Galileo Galilei, *Sidereus nuncius*, 1610. Image courtesy of History of Science Collections, University of Oklahoma Libraries; Copyright the Board of Regents of the University of Oklahoma.

Galileo's drawing of the Moon printed in his *Sidereus nuncius*, or *Starry Messenger*, of 1610 is not accurate. Neither the huge round crater on the terminator, the nearby field of smaller craters on the bright side of the terminator, nor the three bright arms crossing the terminator into the dark hemisphere are seen in modern photographs.

Perhaps in this small engraving Galileo had to depict a crater much larger than it actually is in order to show the left side of the crater illuminated and the right side in shadow (as it would appear with sunlight coming from the right side of the drawing). Or, perhaps Galileo's drawing is an indication of the psychological impact on his thinking of Albategnius, an actual and considerably smaller crater in this approximate location. Or, perhaps Galileo's vision and imagination were enhanced by his expectations.

The Englishman John Harriot (1560–1621), who accompanied Sir Walter Raleigh (ca. 1551–1618) to Virginia in 1585 as cartographer and navigator and published *A Briefe and True Report of the New Found Land of Virginia* three years later, observed the Moon with a telescope a few months before Galileo did, on July 26, 1609. Correspondence between Harriot and his friends reveals their initial confusion: they had no idea that they were seeing shadows cast by craters and mountains. Harriot's drawing is shaded haphazardly, contains no suggestion of craters or mountains, and has the terminator curving in an impossible manner. Early telescopes rarely were powerful enough to enable naive unbiased observers to recognize what they were seeing.

Harriot could have seen a copy of Galileo's book by early July 1610, and that summer he discussed with his friends Galileo's observations. Harriot's July 1610 drawing of the Moon contains the same enormous crater, the same three jagged protrusions, and the same field of craters all found in Galileo's drawing, but not in modern photographs.

In 2007, Galileo's own copy of *Sidereus nuncius*, lost for nearly 400 years, was found in a book collection in South America. On pages eight, nine, and ten of this copy, but not in any other, are five small watercolors in ocher and brown traced by Galileo, a talented painter, highlighting the Moon's craters and valleys.

accompany it in an annual movement about the Sun. Four stars wandered around Jupiter, as did the Moon around the Earth, while all of them together traced out a grand revolution around the Sun.

Galileo named the four objects never before seen the Medicean Stars, after Cosimo II de' Medici (1590–1621), the Grand Duke of Tuscany, and his three brothers. In the introduction to his *Starry Messenger* reporting the discovery, Galileo noted that pyramids and marble or bronze statues constructed to pass down for the memory of posterity names deserving immortality all had perished in the end; the fame of Jupiter, Mars, Mercury, and Hercules, on the other hand, enjoyed, with the stars, eternal life. Now, just as the immortal graces of Cosimo's soul had begun to shine forth on Earth, bright stars offered themselves like tongues to speak of and celebrate his most excellent virtues for all time.

For centuries, but not for all time. Cosimo's Medicean stars are now widely known as the Galilean moons. Furthermore, the German astronomer Simon Marius (1573–1624), who claimed to have discovered Jupiter's moons at the same time that Galileo did, named them Io, Europa, Callisto, and Ganymede, and these individual names have continued in use.

Galileo's contemporaries would not have found his words overly obsequious, in an age when kings and dukes subsidized economically unproductive occupations, including painting, music, literature, and science. Now, democracies often find it difficult to replace royal patronage.

In the spring of 1610, at Easter, Galileo traveled to Florence to demonstrate the use of the telescope and show the new stars to Cosimo. The Medici family had been establishing a classical mythology to justify their rule of Florence, and their palace was filled with frescoes of Jupiter, the supreme Roman god of light and sky, and also protector of the state and its laws. Cosimo responded as Galileo had hoped, and in July invited Galileo to take up residence in Florence as mathematician and philosopher to the grand duke, and also as chief mathematician of the University of Pisa, without obligation to teach.

That fall or winter, Galileo focused an improved telescope on Venus. Were it a planet like the Earth and shining with light borrowed from the Sun, it should show phases, as does the Moon. And it did. Galileo's observations were consistent with the Copernican system, but not the Ptolemaic.

Another telescopic observation with revolutionary implications was of sunspots; these blemishes refuted the ancient idea of an incorruptible and unalterable Sun. From analogy with terrestrial phenomena, Galileo concluded that sunspots were clouds on the Sun. To philosophers who said that the spots were illusions, Galileo replied that these philosophers likewise had become invisible and inaudible; they not so much plumbed the profundity of ancient physics as they conserved the imperious authority of Aristotle.

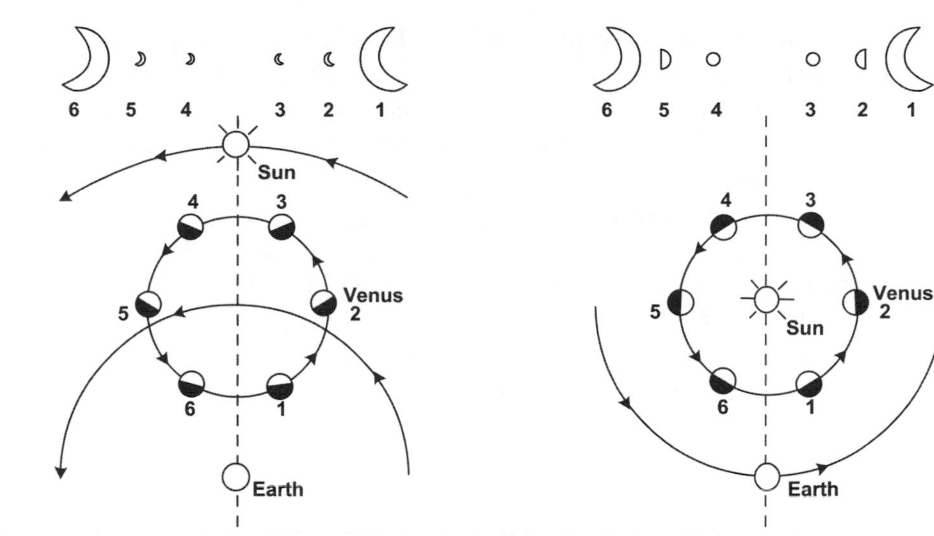

Figure 6.4: Phases of Venus in Ptolemaic and Copernican Models. In both the Ptolemaic geocentric model on the left and the Copernican heliocentric model on the right, Venus appears much larger when it is closest to the Earth, between locations 6 and 1, than when it is farthest from the Earth, between locations 3 and 4. In the Copernican model on the right, Venus, reflecting light from the Sun, exhibits a full range of phases, from new (between 6 and 1) through crescent to full (between 3 and 4). In the Ptolemaic model on the left, however, Venus does not display a full phase to observers on the Earth. From Norriss S. Hetherington, *Planetary Motions. A Historical Perspective* (Westport, CT and London: Greenwood Press, 2006); after Albert van Helden, *Sidereus Nuncius or The Sidereal Messenger. Galileo Galilei. Translated with introduction, conclusion, and notes by Albert van Helden* (Chicago: University of Chicago Press, 1989).

A DIFFERENT UNIVERSE

Above all, the Copernican revolution was a revolution in understanding mankind's place and meaning in the universe. The revolution saw a historical progression from belief in a small universe with humankind at its center to a larger, and eventually infinite, universe with the Earth not in the center. The physical geometry of the universe was transformed from geocentric and homocentric to heliocentric, and eventually to acentric.

All in Doubt

The psychological change was no less than the geometrical change. Mankind no longer commanded unique status, residing in the center of the universe and enjoying this privileged place. Nor was it likely that humans any longer were the only rational beings in the universe. One might even question whether a good God had sent an Adam and an Eve and a Jesus Christ only to the Earth, or to every inhabited planet.

Faith in an anthropocentric universe lay shattered, leaving humankind's relationship with God uncertain, so the English poet John Donne (1572–1631) rhymed in his 1611 poem *The Anatomy of the World*. That the new philosophy called all in doubt and that hierarchical linkages between prince, subject, father, and son were forgotten applied to Christian morality as well as to the physical locations of the Sun and the Earth. Donne wrote:

> And new philosophy calls all in doubt,
> The element of fire is quite put out;
> The Sun is lost, and th' Earth, and no man's wit
> Can well direct him where to look for it.
> And freely men confess that this world's spent,
> When in the planets, and the firmament
> They seek so many new; they see that this
> Is crumbled out again to his atomies.
> 'Tis all in pieces, all coherence gone;
> All just supply, and all relation;
> Prince, subject, father, son, are things forgot,
> ...
> This is the world's condition now, and now
> She that should all parts to reunion bow,
> She that had all magnetic force alone,
> To draw, and fasten sund'red parts in one;
> She whom wise nature had invented then
> When she observ'd that every sort of men
> Did in their voyage in this world's sea stray
> And needed a new compass for their way;
> ...
> She, she is dead; she's dead: when thou know'st this,
> Thou know'st how lame a cripple this world is.

The Great Chain of Being

The idea of a great chain of being, the manifestation in the world of God's thought, would eventually be among things forgotten and abandoned in the intellectual turmoil of the Copernican revolution. The chain had linked God to man to lifeless matter in a world in which every being was related to every other in a continuously graded, hierarchical order. Belief in this hierarchical universe is manifest in the scientific goal of the Royal Society of London in its early years: to follow all the links of the chain of the diverse orders of creatures until all their secrets were uncovered.

Implicit in the idea of a great chain of being was the perfection of the existing state of affairs in social and political hierarchies. The world was not only the most admirable physical mechanism, but it was also the best

The Great Chain of Being

In this engraving, the outermost rings are paradise; then the sphere of the stars; rings for the planets, Sun, and Moon; and the Earth in the center. God reaching out from behind a cloud at the top of the engraving holds a chain binding Nature. She holds a chain binding a monkey, perched on the Earth and representing the physical world. The elements, humans, the arts, plants, and animals all have their assigned places in *Integrae Naturae speculum Artisque imago* (The Mirror of the Whole of Nature and the Image of Art).

This illustration is from a book by Robert Fludd (1574–1637), an English physician, astrologer, and philosopher of the occult. He believed that the human heart was like the Sun and human blood like the planets, and that blood consequently circulated in the human body similarly to the circulation of the planets around the Sun. William Harvey (1578–1657) later explained circulation of the blood in more modern terms, and with experimental evidence.

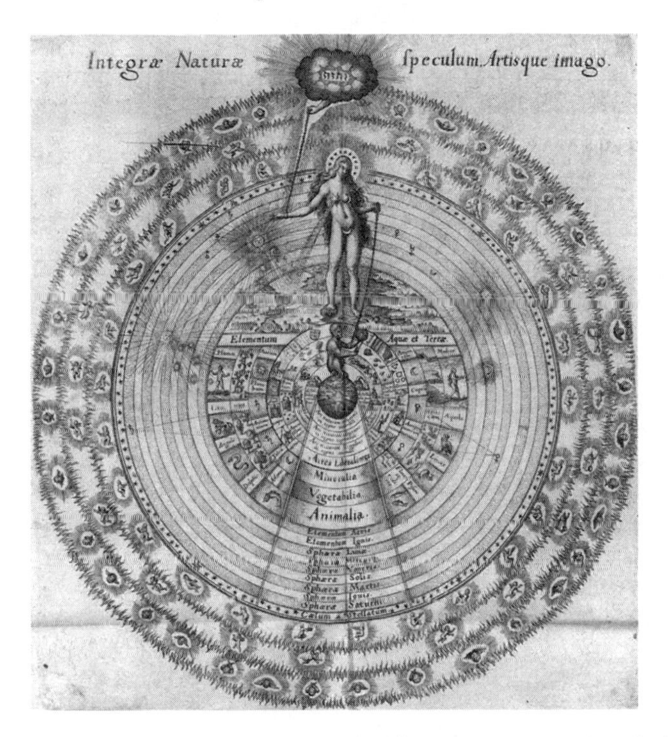

Figure 6.5: The Great Chain of Being. From Robert Fludd, *Urtriusque cosmi maioris scilicet et minoris metaphysica, physica atque technica historia* [The metaphysical, physical, and technical history of the two worlds], 1624. Library of Congress.

political organization, through which the greatest measure of happiness was conferred on its inhabitants. Therefore it was people's moral duty to live appropriately to their places in the hierarchy and not subvert the order of society. Governmental order reflected the order of the cosmos, and belief in the great chain of being precluded the possibility of evolution; social mobility and political change were crimes against nature.

The Protestant reformer John Calvin (1509–1564) acknowledged subtle philosophical comparisons between celestial and earthly hierarchies, but nonetheless rejected papal authority and ecclesiastical hierarchy of church government on Earth. Voltaire (1694–1778), the witty French Enlightenment writer, rejected any comparisons themselves, however subtle and philosophical, between cosmological and church hierarchies. It was pleasing to the imagination, he acknowledged, to contemplate imperceptible ascent from the inanimate to the organic and from plants to zoophytes to animals to angels up to God Himself, and descent from pope through cardinals, archbishops, bishops, curates, vicars, priests, deacons, sub-deacons, and friars to capuchins. But in the ecclesiastical hierarchy, even the humblest member could become pope, while in the cosmological hierarchy even the most perfect creature could not become God.

CONCLUSION

All began to change with the Copernican revolution and destruction of the medieval hierarchal cosmos. Mankind's complacency and sense of self-importance was undermined. It was increasingly unlikely in an increasingly acentric and infinite universe that inhabitants of one tiny speck, the Earth, had already created the best of all possible worlds. Claims of perfection could no longer stifle legitimate social discontent or equalitarian movements. Ideas manifest themselves and have consequences throughout the many arbitrary compartments of human culture. Walls between science, philosophy, religion, literature, art, and politics were breeched. The new astronomy became an integral part of Western civilization and culture.

RECOMMENDED READING

Cohen, I. Bernard. *The Birth of a New Physics, Revised and Updated* (New York: W. W. Norton & Company, 1985).

Drake, Stillman. *Discoveries and Opinions of Galileo*, translated with an introduction and notes by Stillman Drake (Garden City, NY: Doubleday & Co., 1957).

———. *Dialogue Concerning the Two Chief World Systems – Ptolemaic & Copernican*, translated, with revised notes, by Stillman Drake, foreword by Albert Einstein, 2nd ed. (Berkeley: University of California Press, 1967).

Hetherington, Norriss S. *Planetary Motions: A Historical Perspective* (Westport, CT: Greenwood Publishing, 2006).

Lovejoy, Arthur. *The Great Chain of Being: A Study of the History of an Idea* (Cambridge, MA: Harvard University Press, 1936.)

van Helden, Albert. *Sidereus Nuncius or The Sidereal Messenger. Galileo Galilei*, translated with introduction, conclusion, and notes by Albert van Helden (Chicago: University of Chicago Press, 1989).

FILMS

Galileo's Battle for the Heavens. 2002. 120 minutes. Color. NOVA television series, WGBH. See also, http://www.pbs.org/wgbh/nova/galileo/

Science and Belief: From Copernicus to Darwin: Fontenelle's Conversations on the Plurality of Worlds. 1973. 25 minutes. Color. BBC, Open University Productions.

WEB SITES

Cosmic Journey: A History of Scientific Cosmology: http://www.aip.org/history/cosmology.

The Galileo Project. http://galileo.rice.edu/Catalog/NewFiles/digges_leo.html.

History of Science Department, University of Oklahoma, Exhibits Online: http://hsci.cas.ou.edu/exhibits/exhibit.php?exbid=1.

Institute and Museum of the History of Science, Florence, Italy: http://galileo.imss.firenze.it/index.html.

NASA Galileo Mission to Jupiter, Jet Propulsion Laboratory: http://www.jpl.nasa.gov/missions/past/galileo.html.

7

Extraterrestrial Life and Science Fiction

Since the Copernican revolution and the demise of the Aristotelian world view, speculations about extraterrestrial life have flourished. What was formerly heresy is now orthodoxy. In science, organized searches are under way to detect transmissions from intelligent civilizations elsewhere in the universe, while occasional planetary probes seek out rudimentary biological life forms in our own solar system. In science fiction, humankind's place in the universe is being explored by means of thought experiments. Sci-fi is our modern mythology, orienting us in our search for knowledge, developing our image of the universe, and recording our more imaginative attempts to relate the known and the unknown.

PLURALITY OF WORLDS

A plurality of worlds—other planets and other people—is featured in much of science fiction. Belief in a plurality of worlds proliferated when logical implications of the Copernican system, Galileo's telescopic discoveries, and the principle of plenitude, which interpreted any unrealized potential in nature as a restriction of the Creator's power, all came together in the seventeenth century. Revived interest in ancient Greek atomism also argued for many worlds.

Atomism

The ancient Greek philosopher Leucippus (first half of the fifth century BC) and his student Democritus (ca. 460–370 BC) had taught that all matter was made of atoms, and Epicurus (341–270 BC) explained everything in terms of an infinite number of atoms moving and interacting in empty space. Because worlds could be created from these atoms, there was no obstacle to an infinite number of worlds—none, that is, other than Aristotelian philosophy. It came to dominate Western philosophy, resulting in the exile of discussions of atomism and a plurality of worlds from the mainstream of scientific thought during the Middle Ages.

The Greek writer Plutarch (ca. AD 46–127) did discuss the nature of the Moon in his book *Concerning the Face which Appears in the Orb of the Moon*, particularly similarities with the Earth. And he asked whether the Moon was habitable. But his book was not in the atomist tradition.

Atomism was preserved in, of all places, an epic poem, *The Nature of Things* by the Roman philosopher and poet Lucretius (ca. 99–55 BC). The poem was recovered in 1417, during the Italian humanist movement. Not until after publication of an edition in 1563, however, was Lucretius's poem appreciated for its scientific content as well as its poetry, and only then did it renew thoughts of atoms and a plurality of worlds.

John Donne and Henry More

The English poet John Donne (1572–1631), who had lamented in his 1611 poem *The Anatomy of the World* the new Copernican philosophy that called all in doubt, now in his 1624 *Devotions upon Emergent Occasions* lamented that men found reason to conceive not only a plurality of every species in the world, but a plurality of worlds—against which God, Nature, and Reason concurred. Also concurring against a plurality of worlds was the English Neo-Platonic philosopher Henry More (1614–1687), who characterized the idea of an infinity of worlds as a monstrous thing. Yet in his 1646 *Democritus Platonissans, or, An essay upon the infinity of worlds out of Platonick principles* More wrote about the hidden organizing power of atoms creating multiple knots, or terrestrial stars, in the fabric of the heavens:

> knots in th' universal stole
> Of sacred Psyche; which at first was fine,
> Pure, thin, and pervious till hid powers did pull
> Together in severall points and did encline
> The nearer parts in one clod to combine.
> And what is done in the Terrestriall starre
> The same is done in every Orb beside
> Each flaming circle that we see from farre
> Is but a knot in Psyches garment tied.

Psyche, lover of Eros-Cupid, the Greek-Roman god of love, was also thought of as the soul, mind, spirit, breath, life, or invisible animating principle of the universe.

Continuing, More wrote about the knots, or other worlds, in Psyche's garment:

> Whose Centres are the fixed starres on high,
> 'Bout which as their own proper Suns are hurld
> Joves, Earths and Saturns; round on their own axes twurld.

Furthermore, these stars heated frigid planets around them, making life possible:

> their main work is vitall heat t'inspire
> Into the frigid spheres that 'bout them fare,
> Which of themselves quite dead and barren are.
> But by the wakening warmth of kindly dayes,
> And the sweet dewie nights they well declare
> Their seminall virtue in due courses raise
> Long hidden shapes and life, to their great Maker's praise.

Giordano Bruno

In addition to atomism, More's thoughts about a plurality of worlds also had been influenced by the Copernican worldview, by Galileo's telescopic discoveries, and by the writings of Giordano Bruno (1548–1600). Even before Galileo's discoveries, Bruno, an Italian cosmologist, philosopher, and priest, had asserted that the excellence of God was magnified and the greatness of his kingdom made manifest in not one but countless suns, and in not a single earth but in a thousand, indeed in an infinity of worlds. Bruno fell foul of ecclesiastics everywhere he went in Europe, including England. There, from 1583 to 1585, he was an outspoken proponent of Copernicus's belief that the Earth did go round and the heavens stand still. Opponents, one of whom would become Bishop of Oxford, and another Archbishop of Canterbury, responded that it was Bruno's head that did run round, and his brains that did not stand still.

Back in Italy, the Inquisition got hold of Bruno in 1593, and after a lengthy trial, with charges including but by no means limited to blasphemy, immoral conduct, and numerous heresies, among them denying the divinity of Christ and claiming the existence of a plurality of worlds and their eternity, Bruno was tied to a pole and burned in the Campo de' Fiori, a market square in Rome, on February 17, 1600. Galileo's observations and the destruction of the Aristotelian cosmos were still in the future, and Bruno's ideas had little immediate influence.

John Wilkins and Francis Godwin

After Galileo's observations, John Wilkins (1614–1672), a clergyman who helped establish the Royal Society of London and served as its first secretary, argued with more success the case for a plurality of worlds, in his 1638 book *The Discovery of a World in the Moone: or, a Discourse tending to prove, that 'tis probable there may be another habitable World in that Planet.* Although

Figure 7.1: New Worlds. Title page from John Wilkins, combined edition, *The Discovery of a World in the Moone and Discourse Concerning a New Planet*, 1640. Stars are at varying distances above the orbit of Saturn. The Moon is in orbit around the Earth, and the Earth in orbit around the Sun, which is in the center of the system, saying, "I give light, heat, and motion to all." Copernicus on the left offers his heliocentric world. On the right, Galileo offers his telescope. Behind Galileo, Kepler wishes for wings with which to visit the new world. Image courtesy Library of Congress.

there was no direct evidence of lunar inhabitants, Wilkins guessed that there were some inhabitants. Why else would Providence have furnished the Moon with all the conveniences of habitation shared by the Earth? Wilkins followed his 1638 book two years later with *Discourse Concerning a New Planet.*

The first science fiction book written in English was published in 1638. *The Man in the Moone: or A Discourse of a Voyage Thither by Domingo Gonsales, the Speedy Messenger* by Francis Godwin (1562–1633), Bishop of Hereford, was a picaresque satire and lunar voyage fantasy. The protagonist of the story harnessed large migratory birds to a sort of chariot, only to find, to his unspeakable fear and amazement, that this particular species migrated to the Moon. There he learned that Lunars don't have capital punishment, but instead banished those of wicked or imperfect disposition to the Earth, mainly to North America, where the whole native race might be descendants of Lunars, as indicated by a common fondness for tobacco.

Godwin also drew scientific conclusions from his thought experiment, as his story may be characterized. The protagonist observed that in space, when the birds stopped to rest, they did not fall back to the Earth, but it was as if they had no weight. From this he concluded that heavy things do not sink toward their natural place at the center of the universe, as Aristotelian philosophers had believed, but must be drawn by some secret property, much as loadstone draws iron.

The Man in the Moone is remarkable for combining exotic adventure, social commentary, and science. Much of science fiction delivers either thrills or thoughts, but not both, and despite its name largely ignores the science.

Other inspirations for writers, in addition to the Scientific Revolution, included European voyages of discovery and overseas trade. Godwin was a friend of the geographer Richard Hakluyt (ca. 1552–1616), who was born in Hereford and who compiled *The Principal Navigations, Voiages, Traffiques and Discoueries of the English Nation* (in three volumes, 1598, 1599, and 1600).

Although the first edition of Godwin's book was so small that only one copy is known to have survived, more than twenty-five editions were published in four European languages over the next century and a half. The English public was further exposed to Godwin's idea of a world in the Moon in 1706, when his story was transformed into the comic play *Wonders of the Sun* by Thomas D'Urfrey (1653–1723), a companion of King Charles II and a well-known man-about-town.

Godwin's book encouraged Wilkins's sequel, to which Alexander Ross (ca. 1591–1654), master of Southampton Grammar School and chaplain to Charles I, responded with a book of his own, in 1646. The title pretty much told it all: *The New Planet no Planet: or, The Earth no wandring Star: Except in the wandring heads of Galileans. Here Out of the Principles of Divinity, Philosophy, Astronomy, Reason, and Sense, the Earth's immobility is asserted; the true sense of Scripture in this point, cleared; the Fathers and Philosophers vindicated; … and Copernicus his Opinion, as erroneous, ridiculous, and impious, fully refuted.*

Spreading Pluralism

Wilkins had extended God's goodness from the Earth to the Moon. The English poet and playwright John Dryden (1631–1700) extended it farther into space in his 1692 poem *Eleonora*:

> Perhaps a thousand other worlds that lie
> Remote from us, and latent in the sky,
> Are lightened by his beams, and kindly nurs'd.

Persistent rumors that England intended to colonize the Moon further spread the idea of a plurality of worlds and the new Copernican astronomy, although the supposed plan was to send colonists to the empty Moon, not to subjugate indigenous lunar inhabitants. The rumors were perhaps begun by Wilkins's quotation of the German astronomer Johannes Kepler, who had predicted that as the art of flying was developed, the successful nation would transplant a colony to the Moon.

The new Copernican astronomy spread to a wider audience through political and social criticism. The organization of lunar inhabitants became either the model of a perfect society or the reflection of all the vices of the Earth's society. This fictional aspect, not incidentally, furnished writers some protection against outraged monarchs, much as the fiction that scientific theories were merely hypothetical had freed them from theological oversight. The Scottish satirist Samuel Colvil (flourished 1673–1707) described what one might see through the telescope in his 1681 *The Whigs Supplication*:

> If he once level at the Moon,
> Either at midnight or at noon,
> He discovers *Rivers*, *Hills*,
> *Steeples*, *Castles* and *Wind-Mills*,
> *Villages* and *fenced Towns*,
> With *Foussies*, *Bulwarks*, and *great Guns*,
> Cavaliers on horse-back prancing
> Maids about a may-pole dancing
> Men in Taverns wine carousing,
> Beggars by the high-way loafing,
> Soldiers forging ale-house brawlings,
> To be let go without their lawings;
> …
> Young wives old husbands horning,
> *Judges* drunk every morning,

Augmenting law-fruits and divisions,
By *Spanish* and *French* decisions;
Courtiers their aims missing,
Chaplains widow-ladies kissing;
Men to sell their lands itching,
To pay the expenses of their kitching,
Physicians cheating young and old,
Making both buy death with gold.

In much the same spirit, Daniel Defoe (ca. 1661–1731), best known as the author of *Robinson Crusoe*, in 1705 published his own satire on English politics, religion, and trade. In *The consolidator: or, memoirs of sundry transactions from the world in the moon*, Defoe's protagonist ascends to the Moon in a flying machine.

In France, Cyrano de Bergerac (1619–1665) also expressed social criticism through satire, in his posthumously published comical histories of the states and empires of the Moon and of the Sun, with heretical elements censored. During his imaginary lunar visit, Cyrano was tried for heresy, for attempting to persuade the Lunarians that their moon, the Earth, was inhabited. A third volume, a comical history of the states and empires of the stars, was either lost or destroyed.

Bernard Fontenelle

Once the barrier in human imagination against extraterrestrial life was breached by imaginary lunar inhabitants, the concept of a plurality of worlds quickly spread to the planets, and then beyond the solar system to other planets circling other suns. In 1686, the French astronomer, mathematician, and writer Bernard Fontenelle (1657–1757) published his *Entretiens sur la pluralité des mondes* [Conversations on the Plurality of Worlds]. The book was an instant best seller, made it onto the Catholic Index of prohibited books, and continues to be read today.

The Russian Orthodox Church delayed publication of Fontenelle's book in that country for a decade after a translation was ready, until 1740. In 1757, Mikhail Vasilyevich Lomonosov (1711–1765), a scientist, poet, and founder of the Moscow State University, wrote his *Hymn to a Beard* (in Russia only priests were allowed to wear beards):

True it be that all the planets
Resemble ours as earthlike objects.
Be on one of them a long-hair
Priest, or self-appointed prophet:
"By my beard, I swear to you,"
He said, "The Earth is through and through
A lifeless planet; all is bare."

One who remonstrated: "Man lives there."
At the stake they burned him
To punish this free-thinker's sin.

Fontenelle may have been inspired by Wilkins's book on a habitable world in the Moon. It was published in French translation in Fontenelle's hometown in 1655, two years before Fontenelle's birth. Fontenelle described imagined evening promenades in a garden with a lovely young marquise. Naturally their conversation turned to astronomy. On the second evening, the hero explained to his eager and enthusiastic companion that the Moon was "*une terre habitée*," although an absence of atmosphere might change that conclusion. On the third night, he continued with the idea that the

Figure 7.2: Other Worlds. Illustration from Bernard le Bovier de Fontenelle, *Entretiens sur la pluralité des mondes* [Conversations on the Plurality of Worlds], 1686. During an evening promenade in a garden, Fontenelle explains to an eager and enthusiastic lovely young marquise that the Moon is "*une terre habitée*." Image courtesy of History of Science Collections, University of Oklahoma Libraries; Copyright the Board of Regents of the University of Oklahoma.

planets were also inhabited. On the fifth and final night, he argued that the fixed stars were other suns, each giving light to their own worlds. What a romantic flirtation! What a way to woo a woman!

Fontenelle and Feminism

Though not explicitly a feminist, Bernard Fontenelle took seriously women's intellectual ambitions, and origins of the feminist social revolution can be found in the Copernican scientific revolution. Fontenelle's respect for female intellect was unusual for his time. In 1672, the French playwright and actor Molière (1622–1673) in his comedy *Les Femmes Savantes* mocked women for trying to better themselves, for involving themselves in anything other than trivial, mindless pursuits—damned if they did think and damned if they didn't.

Fontenelle helped open a market for women readers, perhaps even more in England than in France. In 1713, an English newspaper described a mother and her daughters making jam while reading Fontenelle aloud to each other. These women were passive consumers of scientific information. In the next advance, women would become active participants in scientific investigations. (See, for example, Madame du Châtelet (1706–1749), in Chapter 9, The Newtonian Revolution.)

Christiaan Huygens

Christiaan Huygens (1629–1695), a Dutch astronomer famous for many things, including a wave theory of light, invention of the pendulum clock, and observations of Saturn's rings and its satellite Titan, wrote *Kosmotheros*: *The celestial worlds discover'd: or, Conjectures concerning the inhabitants, plants and productions of the worlds in the planets.* The book appeared only after Huygens' death, in its original Latin and in English translation in 1698, in French in 1702, in German in 1703, in Russian in 1717, and in Swedish in 1774. Huygens summed up the situation cogently in the opening paragraph of his book, citing the Copernican heliocentric system and observations of satellites of Jupiter and Saturn and mountains and valleys on the Moon:

> A Man that is of *Copernicus*'s Opinion, that this Earth of ours is a Planet, carry'd round and enlighten'd by the Sun, like the rest of them, cannot but sometimes have a fancy, that it's not improbable that the rest of the Planets have their Dress and Furniture, nay and their Inhabitants too as well as this Earth of ours: Especially if he considers the later Discoveries made since *Copernicus*'s time of the Attendents of *Jupiter* and *Saturn*, and the Champain and hilly Countrys in the Moon, which are an Argument of a relation and kin between our Earth and them, as well as a proof of the Truth of that System.

The vicar of Greenwich was so impressed with Huygen's book that he bequeathed £1,902 to Cambridge University to erect an observatory and maintain a professorship of astronomy and experimental philosophy. In

Russia, Peter the Great (1672–1725) commanded translation and publication of Huygen's book there. The publisher, afraid not to do so, but also afraid to publish a work of such Satanic perfidy, compromised by publishing only thirty copies, in 1717. It was the first book published in Russian to describe the Copernican system. Another publisher brought out a larger edition in 1724.

American Almanacs

While belief in extraterrestrial life advanced in European imaginations from the Moon through the planets and on to other solar systems, the concept of a plurality of worlds also spread from Europe to the American colonies. In his *Poor Richard's Almanack*, containing seasonal weather forecasts and practical household hints, and published every year from 1732 to 1758, the American polymath Benjamin Franklin (1706–1790) wrote in the 1749 edition that it was the opinion of all modern philosophers and mathematicians that the planets were habitable worlds. Franklin wondered what sort of physical constitutions the people who lived on Mercury must have, given that the heat of the Sun was seven times greater there than it was on the Earth. Benjamin West, professor of mathematics at Brown University, in the 1778 edition of his *Bickerstaff's Boston Almanack* compared Mercury with Saturn:

> Strange and amazing must the difference be,
> "Twixt this dull planet and bright Mercury;
> Yet reason says, nor can we doubt at all,
> Millions of beings dwell on either ball
> With constitutions fitted for that spot.
> Where Providence, all-wise, has fix'd their lot.

Mythology gave some hint of extraterrestrials' natures. Inhabitants of Mercury were very sprightly, both in movement and in speech, and thus made good lawyers and pettifoggers, at least according to a 1785 almanac. Martians had a warlike disposition. And Venusians were much given to licentious love.

David Rittenhouse (1732–1796), America's best known astronomer, in a 1775 talk before the American Philosophical Society, moved from implications of the Copernican theory to the conclusion that many other things unknown and even inconceivable probably existed in unlimited space. He contended that his conclusion was not in conflict with Christianity. Divine Providence's infinite wisdom and powers might have created throughout the vast extent of creation beings of other rank and degree. These happy people, denied communication with the Earth, were spared Earth's vices and violence. None were doomed to slavery simply because their bodies were disposed to absorb or reflect rays of light in different ways, nor were they subject to the rapacious hand of the haughty Spaniard or the unfeeling British nabob. Thus Rittenhouse commingled pluralism, providence, and patriotism.

Thomas Paine and the *Age of Reason*

Pluralism and providence frequently were linked together, but there was also an inherent tension between them. Thomas Paine (1737–1809), the revolutionary pamphleteer whose 1776 *Common Sense* had advocated independence for the American colonies, wrote in his 1793 *Age of Reason* that to believe that God had created a plurality of worlds rendered the Christian system of faith in the story of Adam and Eve and the apple, and in the story of the death of the Son of God, little and ridiculous, and scattered these stories in the mind like feathers in the air. The two beliefs, in pluralism and in God, could not be held together in the same mind, Paine asserted, and anyone who believed they could be had thought but little of either. Every evidence of the heavens, Paine charged, directly contradicted Christian faith or rendered it absurd.

Although enthusiasm in the United States for Paine's *Age of Reason* was sufficient to justify sixteen more editions in the first four years following its initial publication, and the book was a favorite of enthusiastic deistic societies and college students, Christianity was too widely accepted and deeply felt to be overthrown by the radical rantings of a filthy little atheist, as Paine was later characterized by Theodore Roosevelt (1858–1919), twenty-sixth president of the United States. Pluralism, however, also was widely accepted by the end of the eighteenth century, as is evident in the growing number of pluralist writings over the two-and-a-half centuries following Copernicus. Christians would have to reconcile their religion with belief in extraterrestrial life and hold together these two thoughts simultaneously in their minds.

THE GREAT MOON HOAX

An impressive and also entertaining manifestation of public enthusiasm surrounding the idea of extraterrestrial life is the reception accorded a nineteenth-century satire. Readers then were so exposed to and excited by a plethora of pluralist writings that they did not initially recognize the satire as such. And when it was exposed as fiction, readers branded it a hoax rather than science fiction. It was one of many writings inspired by the idea that there are numerous planets throughout the universe on which intelligent life may exist.

On Tuesday morning, August 25, 1835, readers of the *New York Sun* seemingly were treated to the first installment of an account of great astronomical discoveries lately made by Sir John Herschel, L.L.D. F.R.S. & c. at the Cape of

Good Hope, ostensibly reprinted from a supplement to the *Edinburgh Journal of Science.* John, even more famous than his father, William, discoverer of the planet Uranus, was indeed at the Cape making astronomical observations, from 1834 to 1838. The purported science journal in Edinburgh, however, never existed. The younger Herschel had, reportedly, made the most extraordinary discoveries and affirmatively settled the question whether the Moon was inhabited, and by what order of things. More on this topic would have to wait, though, until after a detailed description of Herschel's telescope. It occupied the remainder of the August 25 installment.

The next day, more than 19,000 copies of the penny daily newspaper were purchased, giving it the largest circulation of any paper on the entire planet. After a long account of the installation of Herschel's telescope at the Cape, the story moved on to observations. Details of the discovery of new stars and nebulae were deferred, so as no longer to withhold from readers the more interesting discoveries made in the lunar world. Lunar vegetation was detected the first night of observations, including trees similar to the largest kind of yews in English churchyards. This demonstrated that the Moon had an atmosphere similar to our own, and thus was capable of sustaining animal life. Lakes and oceans next were viewed, and then animal beings—quadrupeds and birds. By now the Moon was low in the sky, and the increasing refrangibility of her rays prevented any satisfactory protraction of the observers' labors. Their minds also were fatigued with the excitement of their discoveries, so they ended the first night of lunar observations with congratulatory drinks all around.

The next two nights were cloudy and unfavorable to observation, but the following night, further animal discoveries were made, of the most exciting interest to every human being. These would be revealed in graphic language in the next edition of the *Sun.*

On Thursday morning, the paper's readers received more details on lunar vegetation, mountains, and quadrupeds, before moving on to a particular region of the Moon for which Dr. Herschel entertained some singular expectations. (To be continued.)

Friday morning's *Sun* opened with Herschel's telescope now focused on a dark narrow lake, its valley opening upon a plain encircled by a magnificent amphitheater. In addition to quadrupeds and birds, the astronomers sighted on the Moon small parties of winged creatures who, after landing, walked erect. In general symmetry of body and limbs, they were infinitely superior to the orangutan on Earth, and from their impassioned and emphatic gesticulations with hands and arms, it was inferred that they were rational beings.

At this point in the story, highly curious passages containing facts that would be wholly incredible to readers who did not carefully examine the principles and capacity of the telescope with which these marvelous discoveries had been made were omitted. They were to be published later by Dr. Herschel, with certificates from civil and military authorities of the colony and several Episcopal, Wesleyan, and other ministers, all of whom had witnessed the wonders.

Shortly after midnight the next evening, the last veil of mist dissipated and the attention of the astronomers was arrested by such remarkable observational treasures to human knowledge that angels might well desire to win. (To be continued in Saturday morning's edition.)

Another of Dr. Herschel's sagacious theories, revealed Saturday morning, was that the flaming mountain in this region must be a great convenience to dwellers during the long periodic absence of sunlight, and that the region would be well populated. The first object the astronomers saw was a temple of devotion, or perhaps of science, which when consecrated to the Creator was devotion of the loftiest order. Neither at this temple, nor at two other identical temples, did the astronomers observe any visitants, other than flocks of wild doves. What had happened to the devotees of these temples? (To be continued.)

There was no Sunday edition of the *New York Sun*, so readers had to wait until Monday morning for the concluding installment of Sir John's great astronomical discoveries. Immediately in the astronomer's telescope (and in the second sentence of Monday's installment), the winged creatures from the lake, only larger in stature and lighter in color, now appeared near the temples. A universal state of amity among all classes of lunar creatures—no carnivorous or ferocious creatures—gave the astronomers the most refined pleasure and doubly endeared the Moon to them.

Things were going less well on the Earth, where a fire supposedly had broken out at Herschel's observatory. Masons and carpenters were procured from Cape Town, and the whole apparatus was again in operation in about a week.

Not until new moon, however, did the weather prove favorable for a continued series of lunar observations. Only in the final paragraph of the newspaper account was the discovery of the superior species, *Vespertilio-homo*, mentioned. They were of great beauty, scarcely less lovely than representations of angels by imaginative painters, and their social economy seemingly was regulated by laws or ceremonies like those of the winged creatures back at the three temples. Their works of art were more numerous, and of a skill quite incredible. Details would be left for Dr. Herschel's forthcoming authenticated natural history of the Moon.

So concluded the newspaper account, but more was to come. The *Sun* sold 60,000 copies in pamphlet form and the *Mercantile Advertiser* began reprinting the series. The *New York Times* was relatively restrained, calling the discoveries probable and possible, while the *Albany Daily Advertiser* hailed the stupendous discovery, the *Daily Advertiser* said that Sir John had immortalized his name, and the *New Yorker* proclaimed a new era in astronomy and science generally.

The *Journal of Commerce*, however, requested from the *New York Sun* a copy of the document to reprint, and thus learned that none existed. What had happened was that a writer at the *Sun*, after reading an account in the

Edinburgh New Philosophical Journal about ideas for signaling to lunar inhabitants, and also several extravagant books by advocates of a plurality of worlds, had set out to write a satire on the subject. Members of the French Academy of Sciences laughed uproariously and uncontrollably when the account was read to them. Initially Herschel was amused, before a flood of inquiries from people who took the intended satire seriously inconvenienced him greatly.

CANALS ON MARS

Another example of public enthusiasm for the idea of extraterrestrial life, this time on Mars, is found in the furor at the end of the nineteenth century over purported observations of canals on Mars and the dying civilization that must have constructed the canals to move ever-scarcer water around the planet.

The major character in this episode was Percival Lowell (1865–1916). After graduation from Harvard in 1876, where his commencement speech was about the nebular hypothesis and the origin of the solar system, Lowell took the customary grand tour and then went to work in his grandfather's office, handling finances for cotton mills. Shrewd investments soon freed him from business, and he traveled to the Far East, about which he wrote four books. There is a story that while returning from Japan in 1893, Lowell learned of the failing eyesight of the Italian astronomer Giovanni Schiaparelli (1835–1910), who had reported observing *canali*, incorrectly translated into English not as *channels* but as *canals*, on Mars. Schiaparelli's eyesight was indeed deteriorating, but he may not yet have known that it was. Lowell, his eyesight the keenest that a leading ophthalmologist had ever examined, now supposedly decided that it was his manifest destiny to continue Schiaparelli's work.

In 1894, Lowell established an observatory at Flagstaff in the Arizona Territory and began searching for signs of intelligent life on Mars. In 1895, he published his controversial hypothesis in a series of articles in the *Atlantic Monthly* and in the book *Mars.* From visible seasonal changes in Mars' polar cap and in the tint of dark areas, Lowell concluded that the planet had both an atmosphere and water, and thus could support life. Seemingly, there were signs of actual inhabitants: an apparent irrigation system of straight canals, visible to Lowell if not to all observers, radiating from central points.

For Lowell, it was enough that the theory might be true. Professional astronomers, however, envisioned alternative explanations, and returned a verdict of not proven. They also objected that Lowell had gone directly from the lecture hall to his observatory to establish his pre-observational beliefs. Scientists would criticize even more fiercely Lowell's 1908 book *Mars as the Abode of Life*, as fancy foisted upon a trusting public by this mysterious watcher of the stars, whose scientific theories, like Edgar Allan Poe's vision of the raven, had taken shape at midnight.

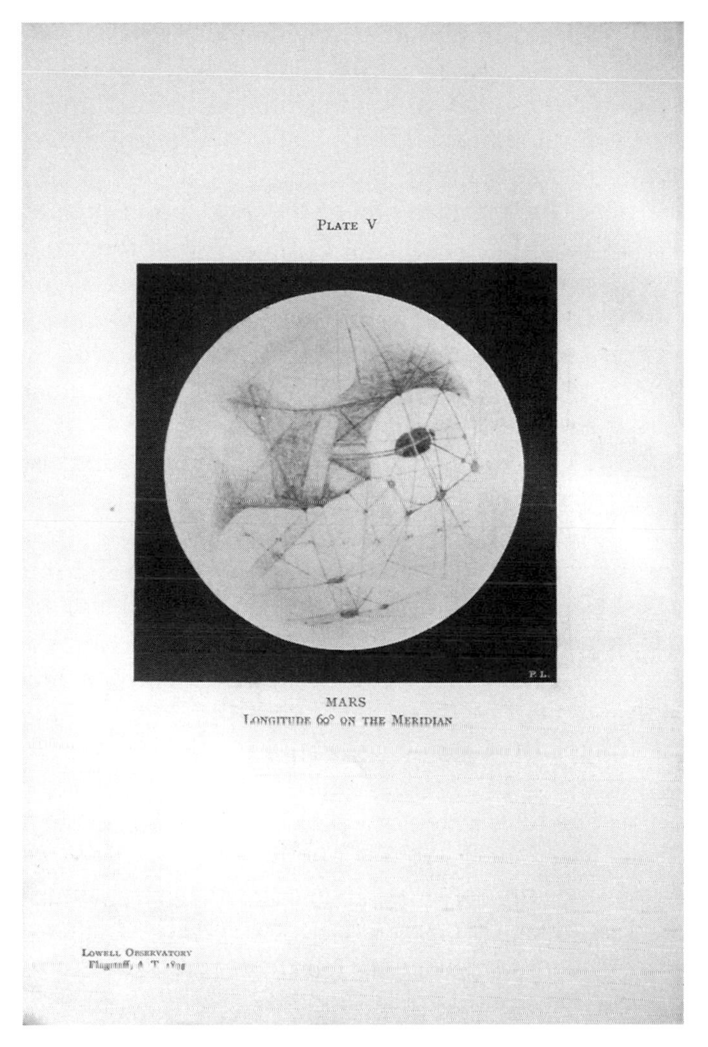

Figure 7.3: Martian Markings. From Percival Lowell, *Mars*, 1895. It was impossible to photograph minute planetary markings directly, so Lowell made drawings of what he observed through the telescope, plotted details from the drawings onto a globe, and then photographed the globe from different angles. In this photograph, Lowell saw a great oval patch 540 miles long and divided by bright causeways, a canal between double dark lines connecting the oval with a long rectangular area, and surrounding the oval a cordon of canals with lakes strung along them like beads on strings. Image courtesy of History of Science Collections, University of Oklahoma Libraries; Copyright the Board of Regents of the University of Oklahoma.

Many readers, however, responded enthusiastically to Lowell's entertaining and informative prose. His Mars was a great red star that rises at sunset through the haze about the eastern horizon, and then, mounting higher with the deepening night, blazes forth against the dark background of space with a splendor that outshines Sirius and rivals the giant Jupiter himself.

Lowell was a poet turned physicist applying his New England intellectual heritage to science. In emphasizing that something was possible rather than that it was not proved, and in legitimizing inferences and intelligently anticipating new developments, Lowell replaced sane but unsensational astronomy with more imaginative thought.

Lowell had taken the popular side of the most popular question afloat. Tenantless globes would be an affront to the sense of the rational in creation, while the discovery of intelligent life on other planets would increase reverence for the Creator. There would also be enlightening social results, because cooperation was seemingly a distinctive feature of life on Mars, where the civilization did not have wars. Mars, whose inhabitants had developed earlier and thus further than had humans on Earth, revealed the future of the Earth: an advanced science and technology, and also an ebbing of life-sustaining resources.

Lowell's purported observations of canals on Mars are now discredited. Partly the product of a psychological inclination to connect minute details too small to be separately and distinctly defined, they are also an instance of preconception influencing perception.

But Lowell's work was not without lasting influence on literature. The American author Edgar Rice Burroughs (1875–1950), best known as the creator of the jungle hero Tarzan, wrote about earthly adventurers traveling to the dry planet Barsoom, where abandoned cities lined former coastlines and scarce water was distributed by canals. The English writer H. G. Wells (1866–1946) had Martians fleeing their dying planet in his 1898 *The War of the Worlds.* There was no immediate panic in the Surrey countryside when an unidentified flying object crashed, nor when creatures emerged from the impact crater. But then they assembled their weapons of death, and the gigantic killing machines marched on London, sweeping humanity aside. It seemed that nothing could save the Earth from the Martian terror.

Forty years later the terror was back, when the American actor, director, and producer Orson Welles (1915–1985) terrified listeners, particularly those who had missed the opening credits, with his adaptation of *The War of the Worlds* as a radio news broadcast reporting the landing of Martians in Grover's Mill, New Jersey. As with the 1835 Moon hoax, intended entertainment was misperceived as reality by a public conditioned to believe in extraterrestrial life, and in 1938, also wrought by anxiety preceding World War II.

In the *War of the Worlds* television series, which ran from 1988 to 1990, Welles was depicted as a government hireling whose broadcast covered up a Martian reconnaissance mission, which preceded war fifteen years later between Earth and Mars. A few conspiracy theorists have imagined that the broadcast was a psychological warfare experiment funded by the Rockefeller Foundation to study the ensuing panic. In a 1992 television episode of *The Simpsons*, Homer gave his son Bart a prank microphone that could tap into local radio systems and thus broadcast on nearby radios, and with the microphone Bart pretended to be the leader of a Martian invasion.

The WAR of the WORLDS
By H. G. Wells
Author of "Under the Knife," "The Time Machine," etc.

Figure 7.4: Martians. Title page from August 1927 *Amazing Stories* magazine featuring H. G. Wells' "The War of the Worlds."

Science, as well as fiction, was drawn into the clamor over life on Mars. The Serbian-American electrical engineer Nikola Tesla (1856–1943) reported from Colorado Springs in 1899 that he had detected coherent radio signals originating from Mars. Guglielmo Marconi (1874–1937), the Italian inventor of radiotelegraphy, was more circumspect; when asked if he had ever heard signals from Mars, Marconi always answered that he was concerned with business on Earth. In 1920, Thomas Edison (1847–1931), the American inventor, commenting on a report that Marconi had said that untraced wireless calls might have come from Mars, confirmed that it was possible. Existing machinery was capable of sending signals to Mars, and if Martians were as far ahead of humans as humans were ahead of chimpanzees, as some said, then the Martians must have instruments delicate enough to hear us, so Edison reasoned.

ADVANCES IN SCIENCE

Changes in astronomical knowledge during the twentieth century eliminated the Moon and Mars as possible abodes of intelligent life, but strengthened the probability of extraterrestrial life elsewhere in the universe. Larger telescopes have expanded the observable universe to millions of galaxies, each containing millions of stars, all rendering it increasingly improbable that Earth alone shelters life.

In 1953, the American chemist Stanley Miller (1930–2007), then a graduate student at the University of Chicago, and his teacher Harold Urey (1893–1981) sent electric sparks through mixtures of methane, ammonia, hydrogen, and water—the supposed constituents of our primitive Earth's early atmosphere—and sparked chemical reactions that produced amino acids, the building blocks of life. Most scientists now suppose that life occurs inevitably on earthlike planets. The television series *Star Trek,* from 1966 to 1969, and four subsequent iterations (*Next Generation*, 1987–1994; *Deep Space Nine*, 1993–1999; *Voyager*, 1995–2001; and *Enterprise*, 2001–2005) visited some of these planets on a weekly schedule.

Skeptics object that if intelligent life is inevitable and has had billions of years to evolve and travel through the universe, it should long ago have reached Earth. That extraterrestrials are not known thus argues against their existence, although believers in UFOs (unidentified flying objects) attribute an apparent, but not real, absence of evidence of extraterrestrials to government cover-ups.

The chemical theory of the origin of life on Earth coincided with the space age, and did not long remain earthbound. In 1976, in one of the greatest exploratory adventures of the twentieth century, and at a cost of over a billion dollars, the National Aeronautics and Space Administration (NASA) landed two Viking spacecraft on the surface of Mars. Experiments detected metabolic activity, but probably from chemical rather than biological processes. In 2005, the European Space Agency's Mars Express probe detected a suggestion of water beneath the Martian surface, water in which primitive microorganisms could have developed over billions of years and might survive today.

After Viking, interest shifted from microorganisms to direct communication with interstellar intelligence. The most comprehensive interstellar communication program was NASA's Search for Extraterrestrial Intelligence (SETI). From a small and inexpensive research and development project during the 1980s, SETI emerged in the early 1990s as a hundred-million-dollar program. A targeted search for radio signals focused on some thousand nearby stars, while a second element of SETI surveyed the entire sky. The program was ridiculed as "The Great Martian Chase," even after changing its name to "High Resolution Microwave Survey" in a vain attempt to highlight its potential for basic discoveries in astronomy, and in 1993, SETI lost its government funding. A scaled-back version of the original targeted search continues with private funding and many volunteers.

SETI had enjoyed a temporary reprieve from congressional budget cutters after publication in 1985 of the science fiction novel *Contact* by Carl Sagan (1934–1996), the preeminent voice of American space science and a national celebrity. In this book, the heroine successfully communicates with alien intelligences.

Earlier, for the 1972 and 1973 Pioneer 10 and 11 spacecrafts sent on flybys of outer planets, and eventually the first spacecraft to escape from the solar system, Sagan and his wife designed plaques with a pictorial message showing the location of the Earth and naked male and female figures. Also in 1973, Sagan's book *The Cosmic Connection*, on the origin of life, extraterrestrial life, and space travel, was published.

The next spacecraft to escape the solar system, two Voyager spacecraft launched in 1977 on flybys of outer planets and their moons, carried golden records with sounds and images portraying the diversity of life and culture on Earth. NASA had been criticized for sending depictions of nudity on Pioneer 10 and 11, and this time humans were depicted only in silhouette.

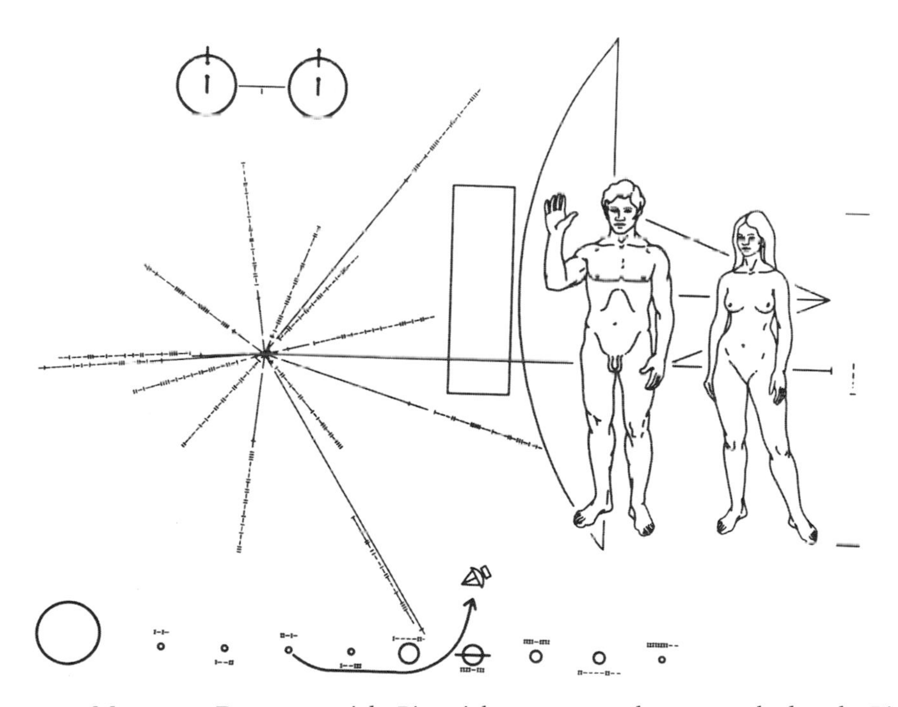

Figure 7.5: Message to Extraterrestrials. Pictorial message on plaques attached to the Pioneer 10 and 11 spacecraft. Male and female figures are drawn to the same scale as the spacecraft behind them. On the left, the position of the Sun is indicated relative to nearby pulsars and the center of our galaxy. At the bottom, our solar system and the planet from which the spacecraft was launched are marked. At upper left is a schematic of the two fundamental states of the hydrogen atom.

Louis Armstrong's *Melancholy Blues* and Chuck Berry's rendition of *Johnny B. Goode* were included, along with excerpts from Johan Sebastian Bach's *Brandenburg Concerto No. 2*, Wolfgang Amadeus Mozart's *Magic Flute*, and Ludwig van Beethoven's *String Quartet No. 13*. The satirical television comedy *Saturday Night Live* announced the first response from extraterrestrials, purportedly a request for more Chuck Berry.

SCIENCE FICTION

Science fiction usually takes the form of exotic adventure or social commentary, often with little astronomical science. Indeed, science can get in the way of science fiction.

Einstein's theory of relativity limits space travel to speeds less than the speed of light. Therefore, the very vastness of space, while offering the enticing possibility of numerous species on zillions of planets in an infinite universe, also makes it very difficult for Earthlings ever to visit many of these remote worlds and meet their alien inhabitants.

A human in the course of a normal life span might possibly travel as far as the Sun's nearest neighbor, Alpha Centauri, 4.35 light years away (the distance a person traveling at the speed of light would traverse in 4.35 years). A trip to the edge of our galaxy, however, would require a minimum of hundreds of thousands of years, and beyond to other galaxies, millions of years, even at the top theoretical speed.

There may be ways around Einstein's relativistic limit, such as a shortcut through a worm hole to another region of warped space. Also, according to relativity theory, travelers experiencing acceleration will age more slowly, albeit not in their own time frame but in time measurements of the civilization they leave behind. Such travelers could conceivably reach a distant planet in a faraway galaxy and return, only to find that in the meantime their home civilization had long since died away.

In the television series *Star Trek*, warp drive propels spacecraft at multiples of the speed of light. Also, transporters convert people into energy patterns to be beamed down to planetary surfaces (much less expensively than landings, whose depiction would have required additional scenery and staging).

In *Dune* by the American science fiction author Frank Herbert (1920–1986), spaceships traverse a folded space, going from one point to another while skipping the space in between. In Sagan's *Contact*, spaceships take shortcuts through wormholes. And in *Foundation* by Isaac Asimov (1920–1992), the incredibly prolific (more than five hundred books) Russian-born American author of popular science and science fiction books, and also a professor of biochemistry, a spaceship can traverse the length of the Galaxy through hyperspace, an unimaginable region neither space nor time, neither matter nor energy, neither something nor nothing, in the interval between two neighboring instants of time.

The British science fiction writer and futurist Arthur C. Clarke (1917–2008) chose to bind himself by the limitations of light speed travel, and in *Childhood's End* he allowed forty years for his travelers to travel forty light-years at 99 percent of the speed of light. Time, however, slows down at that speed, and the passengers only aged two months during their voyage.

Adventure doesn't require space travel. Its spirit is the same, whether in a balloon in *Around the World in Eighty Days* by the French writer Jules Verne or in a submarine in his *Twenty Thousand Leagues under the Sea*. Even when the action takes place off the Earth, such as in Edgar Rice Burroughs's Barsoom series or in the Darkover series by Marion Zimmer Bradley (1930–1999), spaceships are either unnecessary or unimportant.

Sometimes protagonists travel in spaceships to faraway planets, but an entire novel can take place on a fictional planet or even on the Earth itself in an imaginary future. Both provide opportunity for social commentary. In *Star Trek: The Next Generation*, the Borg are found faraway on gigantic cube-shaped spaceships, in which everyone is part machine and operates, powerfully and efficiently, but also undesirably, as part of a hive with no allowance for individuality. George Orwell's *1984* presents a similar dystopia on Earth, no space travel required.

Providers of social commentary from a different viewpoint can come to Earth from space, as did Robin Williams in the *Mork and Mindy* television comedy series, or explore social issues in a different cultural context on a different planet, as the American author Ursula Le Guin (b. 1929) does in *The Dispossessed* and in *The Left Hand of Darkness*, in neither of which is space travel essential. One could even comment from another social stratum or country on Earth.

The idea of a plurality of worlds, of numerous planets inhabited by intelligent life, is so fascinating that it has spawned numerous books, fiction and nonfiction, as well as films, television series, and Web pages. In the *Star Trek* series, a new species was encountered almost every week. And as budgets increased, the physical diversity of alien species became ever more noticeable; Klingons, for example, sported ridges on their foreheads.

CONCLUSION

Would intercourse with extraterrestrials be beneficial? Suppose that they were a cancer of purposeless technological exploitation intent on enslaving Earthlings, rather than benign philosopher kings eager to share their wisdom with us? Even were they helpful and benign, superior beings still would be menacing. Anthropological studies of primitive societies confident of their place in the universe find them disintegrating upon contact with an advanced society pursuing different values and ways of life.

RECOMMENDED READING

Crowe, Michael J. *The Extraterrestrial Life Debate 1750–1900: The Idea of a Plurality of Worlds from Kant to Lowell* (Cambridge: Cambridge University Press, 1986).

———. *The Extraterrestrial Life Debate, Antiquity to 1915: A Source Book* (Notre Dame, IN: Notre Dame University Press, 2008).

Dick, Steven J. *Plurality of Worlds: The Extraterrestrial Life Debate from Democritus to Kant* (Cambridge: Cambridge University Press, 1982).

———. *The Biological Universe: The Twentieth-Century Extraterrestrial Life Debate and the Limits of Science* (Cambridge: Cambridge University Press, 1996).

———, and James E. Strick, *The Living Universe: NASA and the Development of Astrobiology* (New Brunswick, NJ: Rutgers University Press, 2004).

Fontenelle, Bernard. *Conversations on the Plurality of Worlds*, translation by H. A. Hargreaves, introduction by Nina Rattner Gelbert (Berkeley: University of California Press, 1990).

Goldsmith, Donald, and Tobias Owen. *The Search for Life in the Universe, third edition* (Sausalito, CA: University Science Books, 2001).

Guthke, Karl S. *The Last Frontier: Imagining Other Worlds, from the Copernican Revolution to Modern Science Fiction*, translated by Helen Atkins (Ithaca, NY: Cornell University Press, 1990).

Johnson, Francis R. *Astronomical Thought in Renaissance England: A Study of the English Scientific Writings from 1500 to 1645* (Baltimore: Johns Hopkins University Press, 1937; New York: Octagon Books, 1968).

Nicolson, Marjorie. *Voyages to the Moon* (New York: Macmillan, 1948).

Shklovskii, I. S., and Carl Sagan. *Intelligent Life in the Universe* (New York: Dell Publishing Company, 1966).

Sullivan, Walter. *We Are Not Alone: The Continuing Search for Extraterrestrial Intelligence, revised edition* (New York: McGraw-Hill, 1993).

WEB SITES

Ashworth, William B., Jr. *The Face of the Moon: Galileo to Apollo: An Exhibition of Rare Books and Maps from the Collection of the Linda Hall Library*: http://www.lhl.lib.mo.us/events_exhib/exhibit/exhibits/moon/index.html.

Fontenelle, Bernard. *Entretiens sur la pluralité des mondes.* Littérature française en édition électronique. http.//www.scribd.com/doc/236572/Fontenelle-B-Entretiens-sur-la-pluralitie-des-mondes.

Huygens, Christiaan. *Kosmotheros. The celestial worlds discover'd: or, Conjectures concerning the inhabitants, plants and productions of the worlds in the planets. Written in Latin by Christianus Huygens, and inscrib'd to his brother Constantine Huygens, late secretary to his Majesty K. William*: http://www.phys.uu.nl/~huygens/cosmotheoros_en.htm.

Lowell, Percival. *Mars*: http://www.bibliomania.com/2/1/69/116/frameset.html.

"The War of the Worlds" by H. G. Wells, as performed by Orson Welles & the Mercury Theatre on the Air, and broadcast on the Columbia Broadcasting System on Sunday, October 30, 1938, from 8:00 to 9:00 P.M.: http://members.aol.com/jeff1070/script.html.

The War of the Worlds Book Cover Collection: http://drzeus.best.vwh.net/wotw/wotw.html.

Wilkins, John. *The Discovery of a World in the Moone*: http://www.gasl.org/refbib/Wilkins__Moon.pdf.

8

Breaking the Circle

The ancient Greek belief in uniform circular motion still dominated Western astronomy two thousand years afterward, even after Copernicus had exchanged the positions of the Sun and the Earth, and Galileo had overthrown Aristotelian physics, but not the Greek geometrical tradition. Observations by Tycho Brahe (1546–1601) interpreted by Johannes Kepler finally broke the circle, but only at the end of a long string of chance occurrences, the absence of any one of which could have resulted in a very different story. Science is rarely done in a vacuum; social and cultural factors are instrumental in the progress of astronomy.

TYCHO BRAHE

The first happenstance of note is the abduction of baby Tycho by his uncle Jörgen Brahe. Tycho's father soon had more sons, and he allowed the childless Jörgen to raise Tycho, with the expectation that Tycho would inherit Jörgen's wealth. The Brahe family was prominent in the governing of Denmark, and Tycho's brothers prepared themselves for government careers. Tycho, however, began studies at the University of Copenhagen in 1559, encouraged by Jörgen's brother-in-law, Peder Oxe (1520–1575), who served as governor of Copenhagen, lord treasurer, and steward of the realm. Tycho became interested in astronomy, and later in a poem he praised the study of the heavens and deprecated the traditional activities of the nobility. Those would have been his too, had he not been abducted as a baby.

In 1562, Tycho traveled to the University of Leipzig with a tutor, who was supposed to keep Tycho focused on studying law. Instead, Tycho observed the heavens while his tutor slept. His observations revealed that astronomical tables were inaccurate, and Tycho resolved to improve them. He also recognized the need for observations night after night with instruments of the highest possible accuracy.

Tycho's Nose

Tycho lost the bridge of his nose in a duel. The two student protagonists had competed in studying mathematics, and the duel may have been over which of them was the more skilled mathematician. Thereafter, Tycho carried a little box of adhesive salve to hold on a nosepiece, supposedly an alloy of gold and silver. However, when Tycho's tomb was opened in 1901, the nasal opening of his skull was found to be rimmed with green, an indication of exposure to copper; perhaps he had a lighter-weight copper nosepiece for everyday wear.

Shortly after Tycho returned to Denmark, in 1566, Jörgen and the king fell off a bridge near the royal castle at Copenhagen (they had probably been drinking). Jörgen did not recover from his icy ducking, but died of pneumonia. He had not yet made Tycho his legal heir, so Jörgen's wife inherited instead. Oxe, now lord high steward of Denmark, arranged for Tycho to be named to the next vacant position at Roskilde Cathedral. Tycho happily anticipated living on the high income of a nobleman and at the same time pursuing a career as an astronomer and scholar—a career barred by custom to noblemen and not usually subsidized by the crown.

Meanwhile, Tycho went abroad again, and in March 1570, in the town of Augsburg in southern Germany, he had the first astronomical instruments of his own design constructed. Paul Hainzel (1527–1581), a wealthy humanist, astronomer, and mayor of Augsburg, paid for construction of a large instrument for measuring the angular height of stars and had it set up on his country estate. The device had a radius of eighteen feet and was so heavy and bulky that forty men were required to put it in place. Accurate but also cumbersome, it required many servants to align it, yet permitted only a single observation nightly. Before he could make many observations, Tycho was called back to Denmark and his dying father.

In Denmark on the night of November 11, 1572, Tycho noticed an unfamiliar, star-like object in the constellation Cassiopeia. Over the next few nights, the object's angle of view did not change with respect to background stars, as it would have were it nearer the Earth than those stars. Evidently, Aristotle had erred when he decreed there could be no change in the heavenly spheres beyond the Moon. As for those who dismissed this implication of Tycho's discovery, he dismissed them as thick wits and blind watchers of the sky. Publication of the discovery made Tycho famous, and he resolved to make astronomy his life's profession.

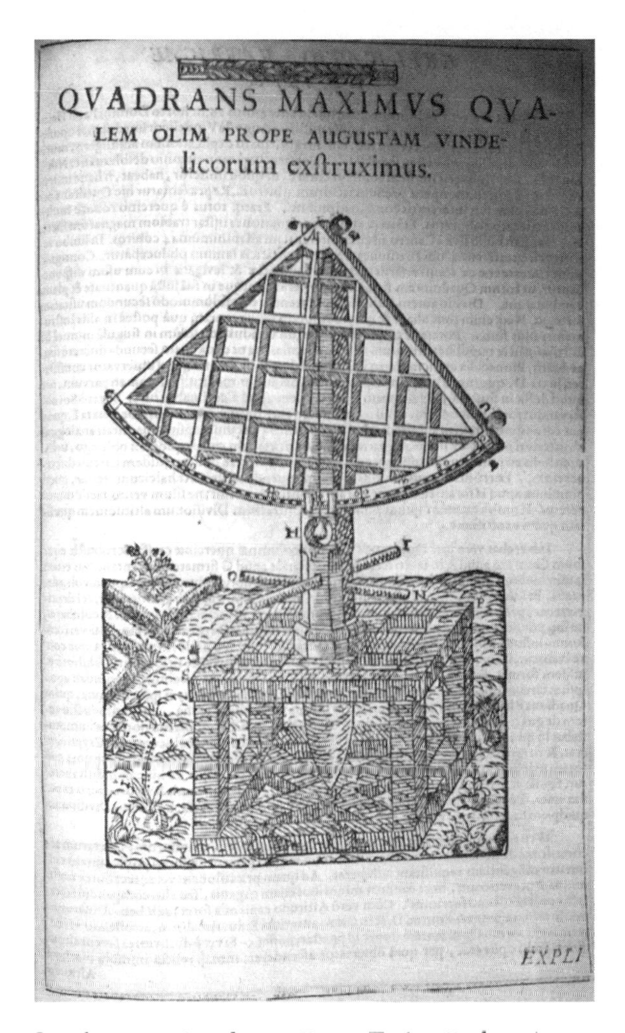

Figure 8.1: Large Quadrant at Augsburg. From Tycho Brahe, *Astronomiæ instauratæ mechanica* [Instruments for the Restoration of Astronomy], 1602. Designed for measuring declination, the angular height of stars, the instrument was constructed of well-seasoned oak, except for the plumb bob suspended from the axis and the brass graduation strip on the ninety-degree arc marked in minutes (sixtieths of a degree). Image courtesy of History of Science Collections, University of Oklahoma Libraries; Copyright the Board of Regents of the University of Oklahoma.

The American poet Edgar Allan Poe (1809–1849) had Tycho's new star in mind when he composed *Al Aaraaf.* (The Qur'an places Al-A'raaf between paradise and hell.) Poe wrote:

> Dim was its little disk, and angel eyes
> Alone could see the phantom in the skies,
> When first Al Aaraaf knew her course to be
> Headlong thitherward o'er the starry sea–
> But when its glory swell'd upon the sky,
> As glowing Beauty's bust beneath man's eye,

We paus'd before the heritage of men,
And thy star trembled—as doth Beauty then!

After the Brahe estate was settled, in 1574, Tycho and his brothers were independently wealthy and could do whatever they wished. While two brothers were contracting marriages, Tycho began planning another trip. He already had a wife, but she was not a noblewoman and they could not be formally married. Also, an aristocratic marriage would have involved Tycho in courtly activities and taken time away from his astronomical studies.

Tycho thought about moving to Basel, on the river Rhine in the Swiss Confederation and far from courtly distractions. However, King Frederick II (1534–1588), eager to retain credit and glory to himself and to Denmark for Tycho's work, offered Tycho the island of Hven and money for building and maintaining a proper establishment there. Generous annual grants amounted to approximately one percent of the kingdom's total revenue. (In comparison, NASA's 1999 budget of $14 billion was 0.8 percent of total federal expenditures and 0.2 percent of gross domestic product.) Tycho also received rents from some forty farms on the island and two workdays per week from each farm. In 1579, he built a manor and observatory with large instruments for observing stellar and planetary positions, a paper mill and printing press, and other necessities for self-sufficiency. A few years later, Tycho added a second, ground-level observatory. It provided a more stable foundation for large instruments, and the placement of instruments and observers in pits in the ground shielded both from chill winter winds.

The new star of 1572 was the first great astronomical event of Tycho's lifetime; the comet of 1577 was the second. He showed that the comet was more distant from the Earth than the Moon and moving through regions of the solar system previously believed to be filled with crystalline spheres carrying the planets around. Aristotle's distinction between the corrupt and changing sublunary world and the perfect immutable heavens was shattered.

Although his observations contradicted Aristotelian cosmology and the medieval worldview, Tycho's world system kept the Earth in the center of the universe. The Sun, the Moon, and the outer sphere of the stars all circled around the Earth. The planets, however, circled the Sun.

In both the Tychonic and Copernican models, the motions of the Sun and planets relative to the Earth are mathematically and observationally equivalent. Also, both systems would accommodate the phases of Venus later observed by Galileo. Tycho gained many of the geometric advantages of the Copernican system without displacing the Earth from the center of the universe or placing it in motion. His arrangement became a popular alternative for astronomers forced by observations to abandon the Ptolemaic scheme but prohibited by religious decree from holding the Copernican theory.

The final chapter in Tycho's life moves him from comfortable Hven to a place where he and Kepler could connect. Tycho became a scandal at the

Uraniborg

Tycho Brahe's great house and observatory building, begun in 1579, was named *Uraniborg*, after Urania, the muse of astronomy and astrology. Supposedly, she was able to foretell the future from the positions of the stars.

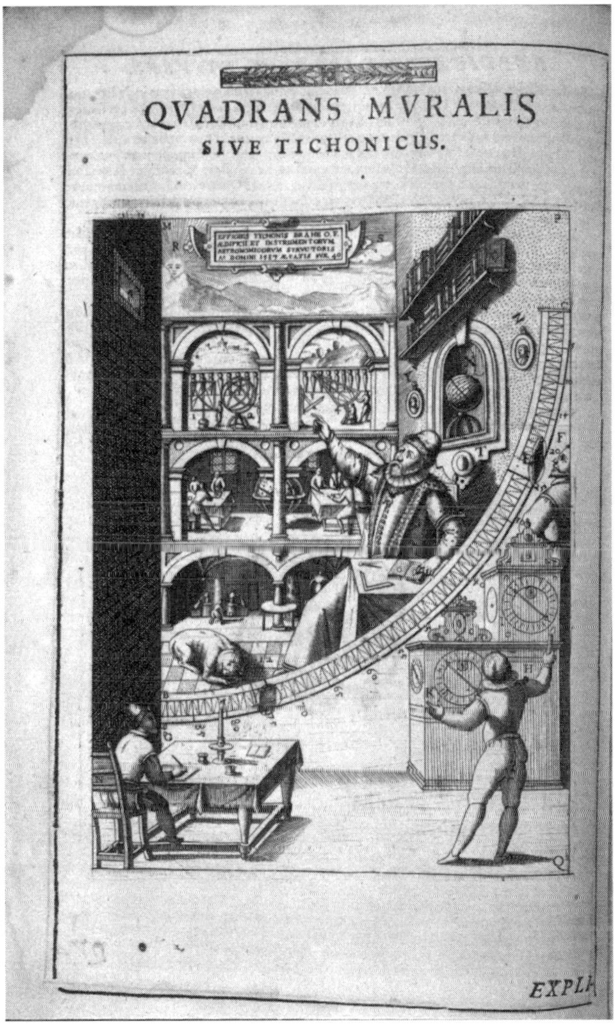

The building, of red brick and carved sandstone trim, formed a large square. Its interior was bisected in each direction by corridors, dividing it into four equal squares. Large towers with removable roof boards were added to the north and south sides of the building, and smaller towers were placed at the front and back, marking the entrances.

All rooms had fireplaces, fourteen in total. The building's front left room was the dining room for Tycho, his family, and some ten students and assistants. In the room behind, a great mural quadrant was mounted to a wall. The adjoining tower contained a great globe and books. Above was the south observatory. Below, in the basement, was an alchemical laboratory. At the other end of the basement were kitchen pantries, cellars, and a well. The kitchen was above, on the main floor. Tycho and his family lived on the floor above. Here, also, was the summer dining room, which was not as well heated as the winter dining room. The towers on this level contained small observing instruments. Above this floor and below the dome, which may have provided light and ventilation, were eight bedrooms for students. Outside galleries surrounded this second level and the top of the house.

Figure 8.2: Great Mural Quadrant. From Tycho Brahe, *Astronomiæ instauratæ mechanica* [Instruments for the Restoration of Astronomy], 1602. Image courtesy of History of Science Collections, University of Oklahoma Libraries; Copyright the Board of Regents of the University of Oklahoma.

In addition to axial symmetry, *Uraniborg* also displayed the numerical progression 1:2:3:4, which contains all the basic Pythagorean musical consonances. The entrance portal was fifteen feet wide and fifteen feet high; the facade of the entrance tower was thirty feet high, with a height of forty-five feet to the peak of its roof; and the building was sixty feet wide.

The great mural quadrant with sights fitted to it for measuring the height of stars was forged of solid brass, six-and-a-half feet in radius, half-an-inch broad, a fifth-of-an-inch thick, and marked in sixths of a minute. The instrument was more impressive in appearance, however, than it was practical.

In an illustration from his 1602 book on astronomical instruments, Tycho points to the heavens. One assistant looks through the back sights, another reads off the time, and a third assistant records the data. The dog at Tycho's feet is a symbol of sagacity and fidelity. In the background is a representation of the main building of Tycho's observatory with an alchemical laboratory in the basement, on the main floor the library with its great globe and space for making calculations, and above, instruments for observing. In a niche in the wall behind Tycho is a small globe, automated to show the daily motion of the Sun and Moon, as well as the phases of the Moon. On either side of the small globe are portraits of King Frederick II and Queen Sophia, whose financial patronage made the whole thing possible. At the top of the engraving is a landscape and the setting Sun.

Stjerneborg

By 1583, Tycho needed more space for more instruments and also a more stable foundation to support larger instruments. A new ground-level observatory, rather than extensions to *Uraniborg*, which would have destroyed its symmetry, was the obvious solution. There would also be no walls, and thus a complete view of the sky.

Excavation began in 1584. Pits terraced with concentric brick circles lowered instruments and observers below winter winds. The terraces also functioned as steps of a ladder, enabling observers to cope with differences of as much as five feet in the heights of the rear sights of Tycho's large instruments, between when the instruments were pointed at the horizon and directly overhead. Wooden roofs over the pits were on wheels and could be turned with little effort.

Tycho named the new observatory *Stjerneborg*, or star town, and had the following words inscribed above the door in gold letters:

> Consecrated to the all-good, great God and Posterity. Tycho Brahe, Son of Otto, who realized that Astronomy, the oldest and most distinguished of all sciences, had indeed been studied for a long time and to a great extent, but still had not obtained sufficient firmness or had been purified of errors, in order to reform it and raise it to perfection, invented and with incredible labor, industry, and expenditure constructed various exact instruments suitable for all kinds of observations of the celestial bodies, and placed them partly in the neighboring castle of Uraniborg, which was built for the same purpose, partly in these subterranean rooms for a more constant and useful application, and recommending, hallowing, and consecrating this very rare and costly treasure to you, you glorious Posterity, who will live for ever and ever, he, who has both begun and finished everything on this island, after erecting this monument, beseeches and adjures you that in honor of the eternal God, creator of the wonderful clockwork of the heavens, and for the propagation of the divine science and for the celebrity of the fatherland, you will constantly preserve it and not let it decay with old age or any other injury or be removed to any other place or in any way be molested, if for no other reason, at any rate out of reverence to the creator's eye, which watches over the universe. Greetings to you who read this and act accordingly. Farewell! [*The Instruments of Tycho Brahe*, frame 28]

Notwithstanding Tycho's plea to posterity, only ruins of *Stjerneborg* now remain.

Figure 8.3: Tychonic, Copernican, and Ptolemaic Systems Compared. From Giovanni Battista Riccioli, *Almagestum novum* [New Almagest], 1651. Riccioli (1598–1671) was a Jesuit and thus prohibited from teaching the Copernican system. Here, Astrea, the goddess of justice, is the winged angel holding the balance beam of truth. The intermediate Tychonic system, with the Moon and Sun circling the Earth, and the planets circling the Sun (though Riccioli has Jupiter and Saturn circling the Earth), outweighs the Copernican heliocentric system. The Ptolemaic geocentric system is discarded in the lower right corner. Image courtesy of History of Science Collections, University of Oklahoma Libraries; Copyright the Board of Regents of the University of Oklahoma.

Danish court for forcing his islanders to work without pay and imprisoning them if they complained. The new king, Christian IV (1577–1648), crowned in 1588, first admonished Tycho for his conduct. Arrogant and domineering, Tycho did not receive admonishment graciously. Christian next slashed

Tycho's funding. In 1596, after a brief stop in Copenhagen, presumably to give Christian a last chance to retain him for the glory of Denmark, Tycho sailed away in search of a new patron, accompanied by his instruments, printing press, and two dozen household servants. After two years of wandering, Tycho found favor with Rudolph II (1552–1612), King of Hungary and of Bohemia, and also the Holy Roman Emperor. Tycho set up his instruments in a castle near Prague. When he died there two years later, in 1601, Kepler would succeed him.

Tycho's Death

According to a note written in Tycho's observation logbook by an assistant, on October 13, 1601, Tycho drank a bit overgenerously at a banquet, felt some pressure on his bladder, but remained seated rather than commit a breach of etiquette. By the time he reached home, he could no longer urinate. After excruciating pain, insomnia, and delirium, Tycho died on October 24.

Presumably, the cause of death was retention of toxic urea in the blood rather than its excretion through urine. Tycho may have been unable to urinate because of an enlarged prostate, although he wasn't all that old. A kidney stone is another possibility, but none were found when Tycho's body was exhumed in 1901.

An examination in 1996 of strands of Tycho's beard found increased levels of mercury. This might be explained by repeated exposure to mercury in his alchemical laboratory while trying to make gold. One of the hairs, however, seemingly had a high local concentration of mercury in the root, suggesting that Tycho ingested the mercury during the last day of his life. If so, did he take it as medicine, by mistake, or was he poisoned?

JOHANNES KEPLER

Kepler's life was every bit as conjoined with cultural factors in a convoluted chain of circumstances as was Tycho's. Born into an impoverished family in southern Germany in 1571, Kepler never would have received an education had not the Dukes of Württemberg created outstanding Protestant universities to produce government administrators and clergymen to lead the reformation then raging in Germany. There was also a network of elementary and secondary schools, and scholarships for children of the faithful poor.

At the University of Tübingen, Kepler received a classical education, which included mathematics and astronomy. Michael Maestlin, Kepler's teacher in these subjects, was a Copernican, and possibly the only openly Copernican professor in the entire world at this time. Seemingly destined for the clergy, Kepler proceeded to the theological school. But when the mathematician at Graz died and the school sought a recommendation from Tübingen for a replacement, Kepler was nominated. Perhaps the Tübingen professors wanted to rid themselves of this querulous young man who had defended Copernicus in a public disputation.

In Graz on July 9, 1595 (so Kepler would later write), he was drawing a figure on a blackboard during a class lecture. Suddenly he was struck so

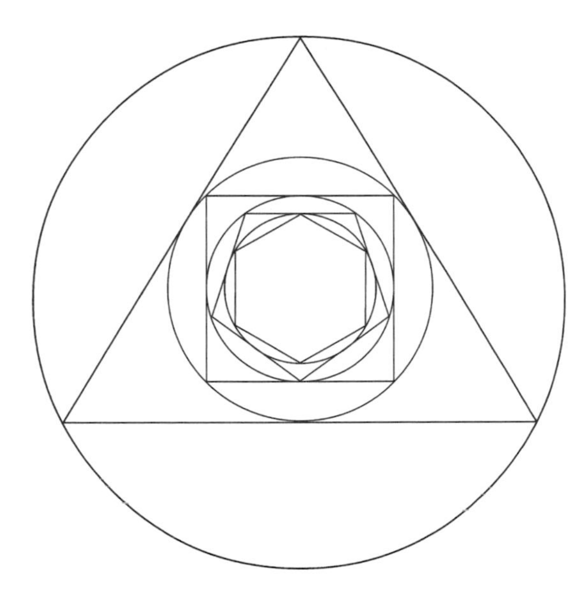

Figure 8.4: Basic Polygons between Circular Orbits. Kepler imagined a mathematical harmony involving planetary radii and polygons. He drew the first basic polygon, an equilateral (all sides equal) triangle, inside the circle of Saturn's orbit. Within this triangle, he drew Jupiter's circular orbit. Remarkably, the ratio between the two circles' radii was the same as that observed between the orbits of Saturn and Jupiter. Then Kepler placed the second basic polygon, a square, inside Jupiter's orbit, and he placed Mars' orbit within the square; then a pentagon (five sides) between the orbits of Mars and Earth; then a hexagon (six sides) between the orbits of Earth and Venus; and a heptagon (seven sides) between the orbits of Venus and Mercury. From Norriss S. Hetherington, *Planetary Motions: A Historical Perspective* (Westport, CT and London: Greenwood Press, 2006).

strongly with an idea that he felt he was holding the key to the secret of the universe. He realized that the ratio of the radii of circles circumscribed around and inscribed within an equilateral triangle is two to one, and that this is the same ratio as that between the radii of the orbits of Saturn and Jupiter. He immediately tried to match the distance between Jupiter and Mars by inscribing a square within Jupiter's orbit and then inscribing Mars' orbit within that square. The match was good enough that Kepler proceeded to inscribe a pentagon within Mars' orbit and inscribe Earth's orbit within that pentagon; then a hexagon within Earth's orbit, and Venus's orbit within that hexagon; and then a heptagon within Venus's orbit, and Mercury's orbit within that heptagon. Kepler thought he had discovered the cosmic order: a grand design linking planetary radii with a progression of basic polygons.

In seeking an understanding of the world in numerical proportions and geometrical figures, Kepler was in tune with ideas and ways of thinking of Renaissance Humanism and Neo-Pythagoreanism. In contrast, Galileo, although he famously stated that nature was written in the language of mathematics, rejected as folly Pythagoras's veneration of numbers and did not seek the structure of the universe in what he derisively called magic numbers.

Kepler's euphoria over basic polygons quickly faded. He realized that there is no natural limit on the number of polygons, and therefore no explanation for the existence of six planets (the number then known), rather than twenty, or a hundred. The need to explain the number of planets, rather than any disagreement between his initial idea and observational data, drove Kepler on. His aesthetic sense required five geometrical figures (corresponding to the five gaps between the six planets) with certain special properties distinct from the rest of the infinite number of geometrical figures. Also, he mused, it would be more appropriate to have solid bodies rather than plane figures between the solid planets.

Kepler thought at once of the five regular solids. The coincidence could not be purely fortuitous, could it? Six planets, five intervals between them, and five regular solids! A frenzied Kepler feverishly devoted his days and nights to computation, to determine whether the resulting astronomical model agreed with observation. One symmetrical solid after the other fit in precisely between the Copernican planetary orbits!

Kepler proclaimed his stupendous discovery in the *Mysterium cosmographicum* [The Secret of the Universe] in 1596, and sent copies of the book to several astronomers. They concluded that Kepler was more a mathematical mystic than a traditional geometrical astronomer. Tycho Brahe, though, did write back, to say that the accurate measurements he had collected over many years could furnish a test of Kepler's theory. Tycho also vaguely invited Kepler to visit him someday (and somewhere, because Tycho was then wandering about Europe in search of a new patron).

Tycho's invitation, however nebulous, was particularly welcome after the Hapsburg archduke decided to cleanse his Austrian provinces of Lutheran heresy. Kepler's school was closed in 1598 and all Lutheran schoolmasters and clergy were ordered to leave in eight days or forfeit their lives. The archduke was so pleased with Kepler's discoveries, however, that he soon allowed Kepler to return. Still, talk of torturing and burning heretics must have been worrisome.

Kepler could not have afforded a trip to distant Denmark, but Tycho was now much closer, in Prague. Kepler would later attribute this proximity to Divine Providence. Furthermore, fortuitous fate found one of Emperor Rudolph's councilors in Graz about to return to Prague, and he agreed to take Kepler with him. Otherwise, Kepler might not have been able to afford even this relatively short trip. They departed for Prague on January 1, 1600.

Kepler also attributed to Divine Providence Tycho's decision to set him to work on the orbit of Mars. The motions of the other planets then known are so close to circular that even Tycho's data would not have revealed their noncircularity; Mars' orbit alone deviates sufficiently from a circle.

Kepler was hired to test Tycho's theories. Not trusting Kepler, Tycho only slowly trickled out observations to him. Tycho's unforeseen death in 1601 enabled Kepler to test his own theory with Tycho's data.

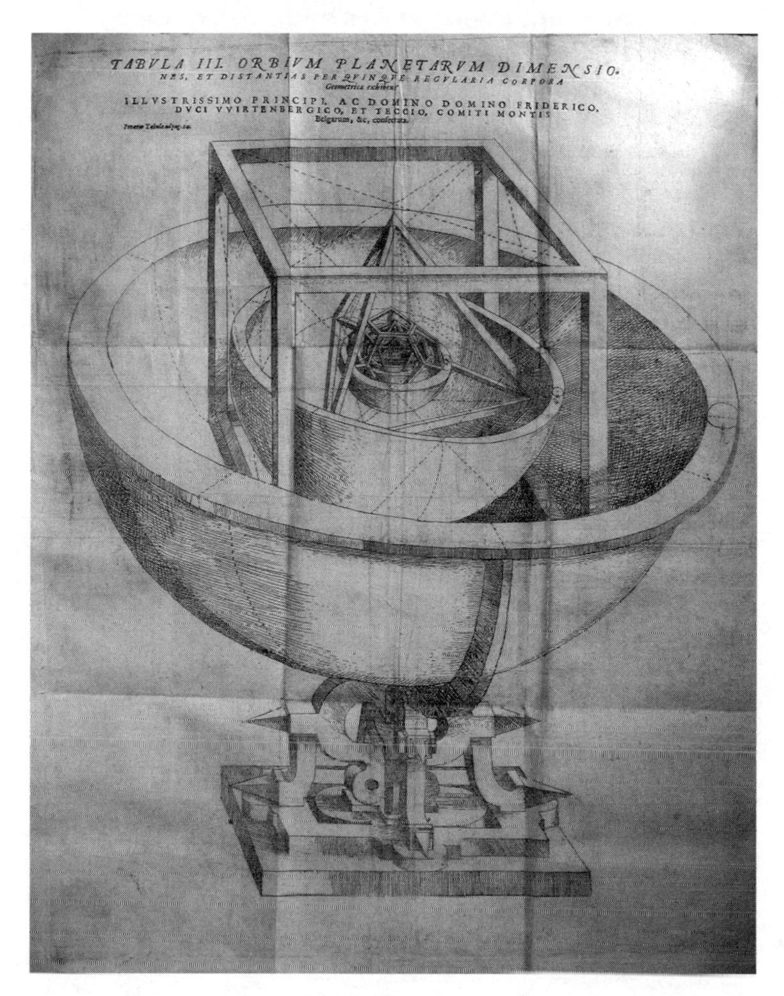

Figure 8.5: Kepler's Model of the Solar System. From his *Mysterium cosmographicum* [The Secret of the Universe], 1596. Saturn's orbit encompasses a cube, which circumscribes Jupiter's orbit, which encompasses a tetrahedron, which circumscribes Mars' orbit, which encompasses a dodecahedron, which circumscribes Earth's orbit, which encompasses an icosahedron, which circumscribes Venus's orbit, which encompasses an octahedron, which encompasses Mercury's orbit. Image courtesy of History of Science Collections, University of Oklahoma Libraries; Copyright the Board of Regents of the University of Oklahoma.

Observations of unprecedented accuracy were crucial, but they were not the only element in Kepler's work. William Gilbert (1544–1603), royal physician to both Elizabeth I (1533–1603) and James I (1566–1625) in England, published *De magnete*, his study of the magnet, in 1600. Now Kepler imagined a force analogous to magnetism, or even magnetism itself, emanating from the Sun and moving the planets. This preoccupation with physical cause, then a proper subject not for astronomers but for physicists, led Kepler to make his calculations with the center of Mars' orbit not at the center of the Earth's orbit but at the Sun, which Kepler believed was both the

Numerical and Geometrical Proportions in Art and the Cosmos

The Humanistic and Neo-Pythagorean fascination with numerical proportions and geometrical figures in nature extended to the microcosm of the human body, as well as to the macrocosm of the universe. The ancient Roman architect Vitruvius (ca. 80–ca. 20 BC) had hoped to find correct proportions for buildings analogous to those of the human body, and his book, recovered by Italian Humanists in 1414, inspired Leonardo da Vinci (1452–1519), the Italian scientist, inventor, and artist, to seek correlations between the human body and the universe. Both possessed an essential symmetry, and geometrical figures—notably the circle and the square—were important for understanding both.

The German engraver and painter Albrecht Dürer (1471–1528) also was concerned with body proportions, which he considered an essential part of Renaissance art theory, and in his engravings *Nemesis* and *Fall of Man* Dürer followed Vitruvius's proportions. Thus Dürer hoped to capture absolute beauty in his engravings, much as Kepler hoped to find beauty and truth in a universe patterned on numerical proportions and geometrical figures—including basic polygons and regular solids. Dürer, too, was mathematically adept, and he discussed geometrical figures in his book on the art of measurement.

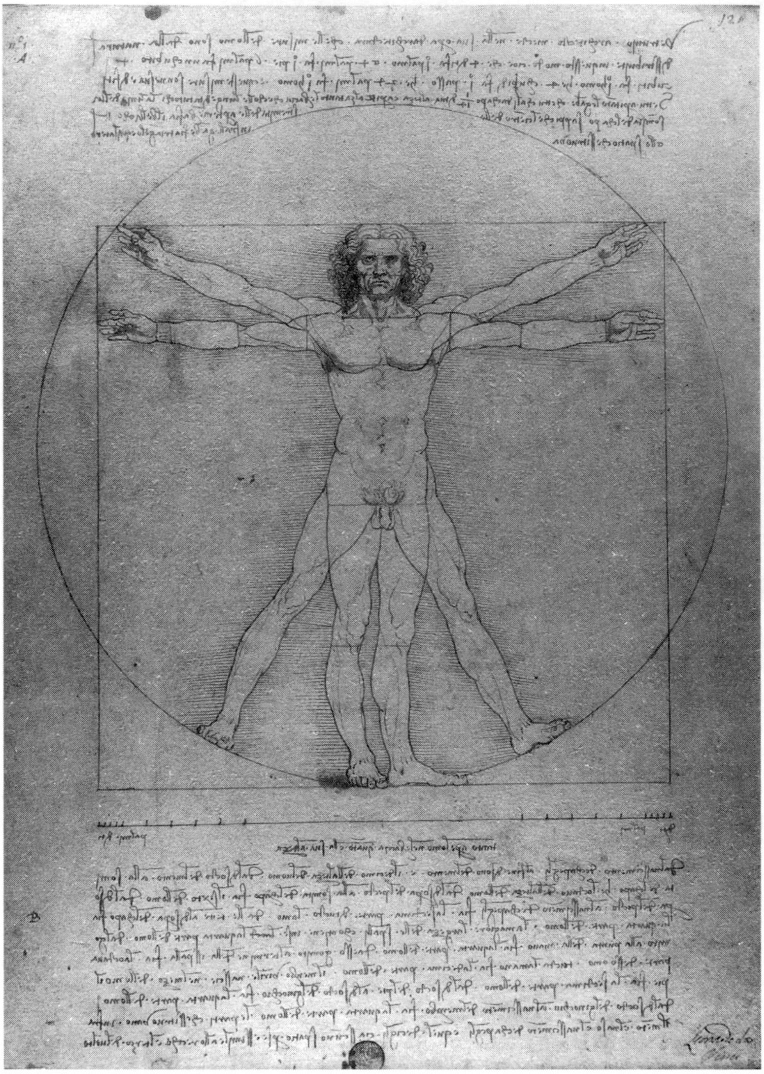

Figure 8.6: Vitruvian Man. Drawing by Leonardo da Vinci, Gallerie dell'Accademia, Venice. Ratios in Vitruvian Man include: length of outspread arms = height; distance from hairline to bottom of chin = one-tenth of height; distance from top of head to bottom of chin = one-eighth of height; maximum width of shoulders = one-quarter of height; distance from elbow to tip of hand = one-fifth of height; distance from elbow to armpit = one-eighth of height; length of hand = one-tenth of height; length of foot = one-seventh of height; distance from bottom of chin to nose = one-third of length of head; distance from hairline to eyebrows = one-third of length of face; length of ear = one-third length of face. Credit: Luc Viatour.

physical center and the cause of the planets' motions. Serendipity smiled, and Kepler obtained a better fit with Tycho's observations.

Kepler tried both circular and oval orbits, and found that discrepancies between them and Tycho's observations were equal and opposite. An elliptical orbit fit in between circular and oval orbits, and better fit Tycho's observations.

Thus Kepler arrived at what would become known as his first law: the planets move in elliptical orbits with the Sun at one focus. He had expected to solve the problem of the orbit of Mars in a few weeks; instead, the solution consumed six years of intense labor.

How could Kepler so casually abandon circular orbits and the accompanying two thousand years of tradition? Maybe because he was more a mathematician than an astronomer. The circle and the ellipse are both conic sections, and Kepler was well acquainted with a book on conic sections by the Greek geometer and astronomer Apollonius of Perga (ca. 262–ca. 190 BC), which had been recovered along with other Greek classics in the early stages of the Renaissance. What is so difficult or revolutionary in substituting one conic section for another? Neither geometrical figure was, for mathematicians, any more natural or more perfect than the other.

Kepler's obsession with physical aspects may have further contributed to the surprising ease with which he abandoned uniform circular motion. Epi cycles might exist in thought, but they weren't real for Kepler.

There was also an immediate practical reason for Kepler to embrace the ellipse. Its geometrical properties simplified his calculations. By now he had discovered another numerical harmony. The line joining a planet to the Sun sweeps out equal areas in equal intervals of time, and it was easier to calculate the area for an elliptical orbit than for an oval orbit.

That the planets move faster the nearer they are to the Sun had already been cited by Copernicus as a celestial harmony. Kepler now found further harmony in a quantitative formulation of the relationship. It was not immediately apparent, however, that this discovery would be hailed enthusiastically in textbooks four centuries later as Kepler's second law.

Arithmetic and geometry had triumphed in the battle against Mars. Now Kepler wanted to press the battle against the other planets. He beseeched the emperor to remember that money is the sinew of war and to order his treasurer to deliver up to General Kepler the money necessary for raising fresh troops. The emperor's treasurer, however, had difficulty even finding funds to proclaim the victory over Mars. Finished in 1606, Kepler's book on Mars, *Astronomia nova* [New Astronomy], wasn't published until 1609.

Kepler's situation worsened. The year 1611 was altogether dismal and calamitous. His salary was in arrears, as usual. War spread to Prague, bringing epidemics. Kepler's wife and one of his children died from disease. And the emperor abdicated.

The following year, Kepler was given the position of provincial mathematicus in Linz, the capital of Upper Austria. At least the Austrians were able to pay his salary. War eventually reached Linz, and in 1628, Kepler moved again, becoming court astrologer to the Duke of Wallenstein. (Astronomy and astrology were not yet distinct activities, one an accepted science and the other a discredited pseudoscience, and astrological duties supported many astronomers financially.) Two years later, while attempting to collect money still owed him by the imperial treasury, Kepler became ill with a fever and died.

The years in Linz had been productive. Kepler published the *Harmonice mundi* [Harmonies of the World] in 1619; the *Epitome astronomiæ Copernicanæ* [Epitome of Copernican Astronomy] in three parts, between 1617 and 1621; and the *Tabulæ Rudophinæ* [Rudolphine Tables] in 1627, finishing the tables of planetary positions which Tycho had begun under the patronage of Rudolph II nearly three decades earlier.

The archetype of the universe remained, for Kepler, the five regular solids. Nor were other harmonic consonances expressed by the Creator neglected, because Kepler believed that God established nothing without geometrical beauty. Kepler even compared the intervals between planets with harmonic ratios in music.

Musical Harmonies

Plato had believed that through imitating the music of the spheres, humankind was returning to paradise. Cicero wrote that music enabled humankind to return to the divine religion. The Italian poet Dante Alighieri (1265–1321) heard the music of the spheres as he ascended from purgatory to heaven. The German composer Johann Sebastian Bach (1685–1750) imagined that the sounds of his organ were a message from heaven. Another German composer, Ludwig van Beethoven (1770–1827), was said to have been contemplating the starry sky and thinking of the music of the spheres when he was inspired to compose the second movement of his String Quartet No. 8 in E minor, opus 59, no. 2. And Einstein wrote that the music of Wolfgang Amadeus Mozart (1756–1791) is so pure that it seems to have been present in the universe waiting to be discovered.

Laws of physics, too, are thought to be preestablished harmonies with stunning symmetries waiting to be plucked out of the cosmos by someone with a sympathetic ear.

That the ratio of the mean movements of two planets is the inverse ratio of the three-halves powers of their spheres is one of many planetary harmonies presented by Kepler in his *Harmonices mundi.* (The relationship, now known as Kepler's third law, is usually stated in slightly different formulations: that the square of the period of time it takes a planet to complete an orbit around the Sun is proportional to the cube of its mean distance from the Sun; or that the ratio of periods squared is equal to the ratio of distances cubed for two planets going around the Sun—and also for two satellites going around a planet.)

Of course, Kepler's initial readers could not have known that this particular relationship would be singled out in the future for great acclaim. More likely, they regarded Kepler's many purported numerical harmonies, including this one, as nonsense.

Kepler might not have discovered his third law, had not John Napier (1550-1617), a Scottish mathematician, astronomer and the eight laird of Merchistoun, invented logarithms and published a description of their use in his 1614 book *Mirifici logarithmorum canonis descriptio* [Description of the Admirable Table of Logarithms]. After Kepler began using logarithms, in 1616, his marvelous mathematical mind would more readily have noticed the correlation between distances squared and periods cubed, because this relationship is more conspicuous as a simple linear correlation between the logarithms of the planets' orbital periods and the logarithms of their distances from the Sun.

In March 1618, Kepler probed the prospect of a correlation between distances squared and periods cubed. But he made errors in his calculations, and failed to corroborate the possibility. Kepler took up the computation again in May 1618, and this time he got it right.

Logarithms were also a godsend, in as much as they transformed the otherwise lengthy process of multiplication into the less time consuming operation of addition. This innovation more or less doubled Kepler's productivity, and also, accordingly, his working lifetime.

Kepler, the Opera(s)

The German composer Paul Hindemith (1895–1963) worked for nearly twenty years before finishing, in 1956, his opera about Kepler: *Die Harmonie der Welt* [The Harmony of the World]. Hindemith shared with ancient Pythagoreans the belief that music is an image of a higher order, that music is number made audible. He was also influenced by the Roman philosopher and mathematician Boethius (ca. AD 480–524), who wrote that humans are subject to the same laws of harmony that govern music and the cosmos, and that we are happiest when we conform to these laws. Hindemith even took the titles for the three movements of his opera about Kepler from Boethius's musical divisions.

Hindemith and Kepler were kindred spirits; both believed that the music of the cosmos caused and ordered the planets' motions and was responsible for the cohesion of the entire universe. They were also victims of political turmoil (Hindemith fled Nazi Germany), and both published philosophical books about their work: Kepler's *Harmonices mundi* and Hindemith's *A Composer's World.* Hindemith recognized in Kepler a person of intelligence and integrity, who in trying to assimilate his life into the harmony of the universe was a model for others.

A second opera about Kepler, titled simply *Kepler*, is scheduled for its premier performance in 2009. The new opera's American composer Phillip Glass (b. 1937) has already written several operas about people with inner visions who altered their cultures. These luminaries include Galileo, Einstein, and the Egyptian pharaoh Akhnaten (see Chapter 2, Mythology).

CONCLUSION

Kepler was the first astronomer to seek universal physical laws based on terrestrial mechanics and comprehending the entire universe quantitatively.

He struggled to develop a physics of celestial phenomena in place of the traditional astronomers' fictitious geometry of eccentrics, epicycles, and equants. Kepler asked questions that others did not, and sought answers where others did not even see problems. Copernicus had noted that the motion of heavenly bodies was circular and rotation was natural to a sphere; thus no further explanation was required. Nor, indeed, was explanation even a concern of Greek geometrical astronomers. They had described how the heavens went, not why. The physical nature and cause of orbital motion was, however, central to Kepler's concerns. And his concerns were a product of the culture of his time.

Kepler's astronomy was revolutionary, or would have been had anyone paid any attention to it. Brilliant as Kepler's work was, it won few adherents during his lifetime. The physical causes, the five regular solids, and the numerical harmonies all were alien to traditional astronomy. Kepler's planetary theories commanded assent because of their unprecedented accuracy, but few astronomers adopted them. There was no Keplerian cosmology or world system, as there were Ptolemaic, Copernican, and Tychonic world systems. Kepler's three laws would be enshrined among the great achievements of science only after Isaac Newton established the concept of universal gravitation.

It is difficult to imagine how astronomical theory might have ended up differently than it has. Yet it is easy to appreciate that its timing and its contorted developmental path owe much to nonscientific factors, especially during the time of Tycho Brahe and Johannes Kepler. Interrelationships between science, society, and culture compellingly and convincingly shaped astronomy.

RECOMMENDED READING

Cohen, I. Bernard. *The Birth of a New Physics, Revised and Updated* (New York: W. W. Norton & Company, 1985).

Hetherington, Norriss S. *Planetary Motions: A Historical Perspective* (Westport, CT: Greenwood Publishing, 2006).

Kepler, Johannes. *The Secret of the Universe-Mysterium cosmographicum*, translated by A. M. Duncan; introduction and commentary by E. J. Aiton; with a preface by I. Bernard Cohen (New York: Abaris Books, 1981).

———. *New Astronomy*, translated by William H. Donahue (Cambridge: Cambridge University Press, 1992).

Koestler, Arthur. *The Sleepwalkers: A History of Man's Changing Vision of the Universe* (New York: Macmillan, 1959). The chapters on Kepler are also published separately as *The Watershed* (Garden City, NY: Doubleday, 1960).

Thoren, Victor E. *The Lord of Uraniborg: A Biography of Tycho Brahe* (Cambridge: Cambridge University Press, 1990).

WEB SITES

The Galileo Project: http://galileo.rice.edu/Catalog/NewFiles/digges—leo.html.
The Instruments of Tycho Brahe: http://www2.kb.dk/elib/lit/dan/brahe/engelsktekst/introframee-en.htm.
The Starry Messenger: http://www.hps.cam.ac.uk/starry/starrymessenger.

9

The Newtonian Revolution

In his parlor at Monticello, Thomas Jefferson (1743–1826), principal author of the Declaration of Independence and the statute of Virginia for religious freedom, third president of the United States, and founder of the University of Virginia, displayed portraits of the three men whom he believed to be the greatest who had ever lived. Prominently in the middle was Isaac Newton (1643–1727). (The Julian calendar, still in use in England during Newton's lifetime, assigned his birth to Christmas Day, December 25, 1642, rather than to January 4, 1643, as does the Gregorian now in use. Furthermore, until 1752 the new year began legally in England on March 25. Therefore the Old Style [OS] date of Newton's death, March 20, 1726, is, in the New Style [NS], March 31, 1727.)

Newton had explained how the planets retrace their paths around the Sun. In doing so, he had provided a new mathematical and physical structure for the universe. For this achievement, Newton was widely acclaimed as the greatest scientist the world had yet known.

Jefferson, however, idolized Newton for more than his scientific achievement. In the eighteenth century, the Age of Enlightenment, many of America's founding fathers were inspired by Newton's discovery of one fundamental law governing the physical world. Now they hoped to extend systematic thinking and reason to all areas of human activity, and to discover the fundamental laws of the human mind, of morality, and of the relationship between individuals, society, and the state.

Much later, in the twentieth century, the British scientist and novelist C. P. Snow (1905–1980) would decry a breakdown of communication between the two cultures of modern society, the sciences and the humanities. But in Jefferson's time, Newton was the primary icon of both cultures.

Developments in astronomy were intertwined with human values, and Western culture was transformed. Reason replaced superstition, progress pushed aside tradition, secularism contested with religion, and science rivaled humanism.

NEWTON'S ASTRONOMY

In ancient Greek and medieval European astronomy, solid crystalline spheres had provided the physical structure of the universe and carried the planets in their motions around the Earth. Then Copernicus removed the Earth from the center of the universe, Tycho shattered the crystalline spheres with his observations of the comet of 1577, and Kepler replaced circular orbits with ellipses. It remained for Newton to show how the planets retraced their paths around the Sun for thousands of years.

Newton's father, although a wealthy farmer and lord of his own manor, could not sign his name. Probably he would not have educated his son; his brother did not. Newton was born three months after his father's death, premature, and so small that he was not expected to survive. When he was three, his mother remarried and Newton was left in his grandmother's care. Difficult early years probably contributed to his difficult mature personality. He rejoined his mother seven years later, after his stepfather died, leaving young Newton's mother wealthy. Her family was educated, and she soon sent Newton off to school. He returned home at age seventeen to learn to manage the family manor. But he was a disaster at rural pursuits, his mind lost in other thoughts. His mother's brother, a clergyman, urged her to send young Newton back to school to prepare for university. Newton entered Trinity College, Cambridge, in 1661. He was appointed a fellow in 1667, and named Lucasian professor in 1669, at age twenty-six. In later years, he would serve as Master of the Royal Mint and President of the Royal Society.

At Cambridge, Newton initially studied Aristotelian physics. But around 1664, his notebooks reveal, he learned of the French scientist René Descartes (1596–1650). Descartes' mechanistic universe consisted of huge whirlpools, or vortices, of cosmic matter. Our solar system was one of many whirlpools, all its planets moving in the same direction in the same plane around a luminous central body. All change in motion was the result of percussion of bodies; one object could act on another only by contact. Celestial matter circulating about the Earth pushed all terrestrial matter toward the Earth. In contrast to Descartes' explanation of the phenomena of nature in terms of particles bouncing off each other, Newton would explain planetary motions as the result of a mysterious attraction at a distance: gravity.

The English scientist Robert Hooke (1635–1703), the astronomer Edmond Halley (1656–1742), and the architect Christopher Wren (1632–1723) all speculated about an attractive force holding the Moon in its orbit around the Earth and the planets in their orbits around the Sun. None of them, however, was able to show what orbit would result from an inverse square force of attraction; none of them could derive Kepler's laws of planetary motion from principles of dynamics.

In London at a meeting of the Royal Society in January 1684, the three men discussed the problem. Halley admitted that he had failed to demonstrate the laws of celestial motion from an inverse square force. Hooke claimed that he could do so. Wren doubted that Hooke could, and offered a prize of a book worth forty shillings to anyone who could produce a demonstration. Hooke replied that he intended to keep his solution secret until others failed, because then they would value his work more.

In August of 1684, Halley visited Newton in Cambridge. While there, Halley asked Newton what he thought the curve would be that would be described by the planets if the force of attraction towards the Sun were the reciprocal to the square of their distances from the Sun. Immediately, Newton replied that it would be an ellipse. Halley asked how he knew this; Newton replied that he had calculated it.

Newton could not, however, find the calculation among his papers. He promised to do the calculation again and to send it to Halley. In November, Halley received from Newton a nine-page essay on the motion of bodies in an orbit. Halley recognized the importance of the work and reported it to the Royal Society.

Meanwhile, back in Cambridge, Newton was extending his geometrical demonstrations to additional phenomena and making the calculations more precise. He was completely absorbed in this work, to the exclusion of virtually everything else. An acquaintance wrote that sometimes Newton even forgot to eat.

The product of Newton's frenzied concentration was the *Principia*, or—to give the book its full title—*Philosophiæ naturalis principia mathematica* [Mathematical Principles of Natural Philosophy]. Again, Halley recognized the worth of Newton's work, and he transmitted the manuscript to the Royal Society. When the Society shirked its duty, Halley undertook to see the book published at his own expense.

Newton's intent, revealed in the title of his book, was to refute Decartes' philosophical principles, which had appeared under the title *Principia philosophiae*, and to replace them with a natural philosophy founded securely on mathematical principles, on the laws of motions and forces.

Newton asserted that if it universally appeared, by astronomical observations, that all bodies about the Earth gravitated toward the Earth in proportion to the quantity of matter that they contained, that the Moon likewise gravitated toward the Earth, and all the planets and comets gravitated toward the Sun, then it must be universally agreed that all bodies were endowed with a principle of mutual gravitation.

After presenting his rules of reasoning, Newton ostensibly began with phenomena. Universal gravitation would come to be understood, rightly or wrongly, as an inductive achievement, the inference of Kepler's laws from observed facts of planetary motions. For each phenomenon, Newton proposed that the forces by which the planets were held in their orbits tended to the Sun and were inversely as the squares of the distances of the planets from the Sun.

Once he knew the principles on which the phenomena depended, from these principles Newton proceeded to deduce the motions of the heavens. From the theory of universal gravitation, all the observed phenomena mathematically followed. Not only Kepler's laws, but many additional phenomena also fell under Newton's geometrical onslaught.

There remained the nature of gravity itself. Newton acknowledged that, while he had explained the phenomena of the heavens by the power of gravity, he had not yet been able to discover the cause of the properties of gravity from phenomena. Now he argued for setting aside the question of what gravity was, and to be content with a mathematical description of its effects. No doubt Newton wanted to explain the cause of gravity, but he wasn't able to do so. He settled for demonstrating that the motions of the celestial bodies could be deduced mathematically from an inverse square force, whatever anyone might imagine the metaphysical or occult qualities of that force to be.

NEWTONIANS VERSUS CARTESIANS

It remained to be seen if Newton's gravity, without physical explanation but increasingly in agreement with observed planetary motions, could win out over Descartes' vortices, more agreeable to the intellect but overwhelmed by technical difficulties. And if gravity did triumph, what repercussions might the victory have beyond astronomy? In hindsight, victory was inevitable. For nearly a century, however, Newtonians and Cartesians battled fiercely.

A propaganda effort on Newton's behalf was already underway even before the *Principia* appeared in print. Halley had written to a few continental scientists informing them of the new theory of universal gravitation brilliantly investigated by Newton, perhaps the greatest geometer ever to exist, which proved how far the human mind properly instructed could succeed in seeking truth. Also before the *Principia* was published, there appeared in the *Philosophical Transactions of the Royal Society of London* (edited by Halley) an anonymous review (written by Halley) of Newton's incomparable treatise, which provided a most notable instance of the extent of the powers of the mind and showed the principles of Natural Philosophy and their consequences.

The French literary giant Voltaire (1694–1778) wittily summed up the situation in 1733 in his *Lettres Anglaises* [Letters Concerning the English Nation]. A Frenchman arriving in London would find philosophy, like everything else, very much changed there. He left the world full, and found it empty; he left the world a plenum, and found it a vacuum. In Paris, the universe was seen composed of vortices of subtle matter; but nothing like vortices were to be seen in London. In Paris everything was explained by a pressure that nobody understood; in London everything was explained by an attraction that nobody understood either.

Royal permission for publication of Voltaire's book had been withheld, and enraged authorities issued a warrant for his arrest. Tipped off by one of the king's ministers, Voltaire had already fled to the comforts of the chateau of Madame du Châtelet (1706–1749), conveniently located, were he pursued, near the border with Lorraine, then an independent province. Left behind in Paris, Voltaire's publisher was thrown into the Bastille, and the public hangman burned copies of the impious book implying that French civilization was deficient. Scandal and the distinction of being banned in Paris fueled sales of the book, which topped 20,000 within five years.

In 1737, Voltaire's new book, *Elêmens de la philosophie de Newton* [Elements of Newton's Philosophy], saw all Paris resounding with Newton, all Paris stammering Newton's name, and all Paris studying and learning Newton's astronomy. Madame du Châtelet, who had assisted Voltaire with this book and had also translated Newton's *Principia* into French, with extensive commentary, wrote to a friend that publication of Voltaire's books was necessary because the French were blissfully ignorant of flaws in Cartesianism evident to everyone else in Europe, and thus unable to participate in the progress that Newton's discoveries would make possible.

As late as 1774, Voltaire's books were still praised as the only source for gentlemen not learned in the sciences to turn to for information about Newton. In England, a few university students, if not cultured gentlemen and ladies, had access to Newton's science as early as 1702, in two new textbooks.

In the same year as the appearance of Voltaire's *Elements of Newton's Philosophy*, the Newtonian revolution also won a major battle on the scientific front. Newton had predicted that the Earth was flattened at the poles, while Descartes' vortex theory asserted that the Earth was flattened at the equator. Voltaire wrote that in Paris the Earth was shaped like a melon, while in London it was flattened on two sides.

Measurements within France by Jacques Cassini (1677–1756), one of four generations of Cassinis to direct the Paris Observatory from its founding in 1669 until they were driven out in 1793 during the French Revolution, tentatively supported the Cartesians. More northerly measurements were required, however, and in 1736, the French Academy of Sciences dispatched an expedition to Lapland. Its members returned a year later as the flatteners of the Earth and of the Cassinis, so Voltaire quipped. Little glory was gained, however, in confirming to Frenchmen a discovery predicted fifty years earlier by an Englishman.

Cartesians were also flattened by repeated failures to reconcile vortex theory quantitatively with Kepler's mathematical laws of planetary motion.

Figure 9.1: Frontispiece, Voltaire's *Elémens de la Philosophie de Neuton*, 1738. Rays of light come from behind Newton, wearing a toga and seated on a throne of clouds. Madame du Châtelet, who had assisted Voltaire with the book, is held in the sky by winged cherubs. She holds a mirror reflecting the light from Newton to Voltaire, busy writing at his earthbound desk. Light was widely employed as a metaphor and as a figure for knowledge. Image courtesy of History of Science Collections, University of Oklahoma Libraries; Copyright the Board of Regents of the University of Oklahoma.

Light

Light has been employed widely over many cultures and for thousands of years as a metaphor and image for knowledge. As early as Plato in fourth-century-BC Greece, in his allegory of the cave, prisoners were freed to the real world of light and knowledge.

The English poet Alexander Pope (1688–1744) eulogized Newton with the couplet:

Nature, and Nature's Laws lay hid in Night.
God said, "Let Newton be!" and all was Light.

To which a twentieth-century British poet added:

It did not last: the devil, shouting "Ho.
Let Einstein be" restored the status quo.

"Let there be Light" was a Biblical announcement of the coming of light and knowledge into the world, and has been adopted, in its Latin form, *Fiat Lux*, as the motto of the University of California.

References to light and to the Sun are a way of saying that people can see and understand the truth. "I see" means "I understand." In legal contexts, one judge found a matter as clear as the noonday Sun when it shines, and another judge wrote that the Sun of truth now shone upon a dark and adulterous intrigue. Sunshine laws require governments to make public their actions, and the *New York Times* topped an article on bureaucratic hindrances to the Freedom of Information Act with the headline "Let the Sun Shine."

The common phrase "in light of" indicates that additional information has been employed to interpret and understand an issue more fully, in the same manner that light from the Sun enables a viewer to see an object more fully.

Neither Descartes nor his early followers had tried to account for Kepler's observational results. Indeed, no one prior to Newton had regarded Kepler's laws as facts requiring explanation by physical theory. But after Newton, Kepler's results had to be explained, and the inability to do so with vortices encouraged astronomers to declare publicly their allegiance to Newton and to gravity.

Through the eighteenth century, Newtonians added to their successes and solidified their victory. The solar system contains many planets, and the calculation of any one planetary orbit is not simply a matter of the gravitational attraction between that planet and the Sun. There are smaller, but not negligible, perturbational effects from the attraction of other bodies on any particular planet. For example, Jupiter and Saturn modify the motions of each other about the Sun, and the Sun alters the Moon's motion around the Earth.

In addition to perturbations of planets and the shape of the Earth, the motions of comets, precession (a slow conical motion of the Earth's axis of rotation caused primarily by the gravitational pull of the Sun and the Moon on the Earth's equatorial bulge), and nutation (a smaller wobble superimposed on the precessional motion of the Earth's axis) were also grist for the eighteenth-century's mathematical mill. In 1785 and 1787, the French mathematical astronomer Pierre-Simon Laplace (1749–1827) seemingly resolved the last major unexplained anomalies in the solar system: a large anomaly in the motions of Jupiter and Saturn and an acceleration of the Moon's orbital speed around the Earth.

CULTURAL REPERCUSSIONS

Laplace also took up the question of the origin of the solar system. Reflecting the new atheistic approach to nature of many scientists of the French Enlightenment, he attempted to replace the hypothesis of God's rule with a purely physical theory that could also explain the observed order of the universe, particularly the remarkable arrangement of the solar system. He was successful, at least in his own mind. According to legend, when Napoleon asked Laplace whether he had left any place for the Creator, Laplace replied that he had no need of such a hypothesis.

Replacement of God's rule with a purely physical theory began the separation of science and religion, previously joined in Western thought. Theology had reigned as king of the disciplines, autonomous, the supreme principle by which all else was understood, its fundamental postulates and principles derived from divine revelation, interpreted and formulated within the tradition, and producing knowledge of ultimate value. Science had been merely a handmaiden, neither controlling fundamental knowledge nor ways of getting at it, its truths holding a lower logical status and value.

Now, flush with triumphant reductions of all known phenomena of the solar system to the universal law of gravity, science began to displace religion as the source to which people turned for inspiration, direction, and criteria of truth. In the future, religion and politics increasingly would appeal to science for legitimacy. (For more on this topic, see Chapter 10, Astronomy and Religion.)

The example of Newtonian astronomy transformed virtually all aspects of Western culture. Having discovered more of the essential core of human knowledge than anyone before him, Newton now illustrated the amazing ability of human reason. Through his great achievement, Newton encouraged others to apply reason to other subjects. The eighteenth century became the century of enlightenment, in which critical human reason would free people from ignorance, from prejudices, and from unexamined authority.

Political thinkers grew increasingly confident that they could determine the natural laws governing human association, and the American and French Revolutions followed. The French social commentator and political thinker Charles Montesquieu (1689–1755) applied the spirit of scientific inquiry to the study of man in society. He noted similarities between scientific laws and social laws, and he hoped to find a few fundamental principles of politics. His 1748 *De l'esprit des lois* [The Spirit of Laws] was banned by the Catholic Church in 1751 and placed on the Index of Prohibited Books. Nonetheless, or perhaps because of this ban, Montesquieu became the most frequently quoted writer on government and politics in prerevolutionary North America.

The British political philosopher Edmund Burke (1729–1797) insisted that reason should govern political decisions and was sympathetic toward the American colonists. Later, however, Burke was dismayed by the violence and chaos in France during its revolution, which he blamed on the undiscriminating application of the rules of reason. Thomas Jefferson, who believed that the Newtonian system of philosophy was as much a part of the common law as was the Christian religion, invoked laws of nature in the

Declaration of Independence. James Madison (1751–1836), in the debates of the Constitutional Convention, called upon astronomical imagery even more explicitly, describing a need for the central government to control the centrifugal tendencies of the states to fly out of their proper orbits and destroy the whole harmony of the political system.

Benjamin Franklin so idolized Newton that he had his own portrait painted sitting at a desk, supposedly inspired by a large bust of Newton towering over him. It was not uncommon to include a bust of Newton in commissioned portraits, or a book in the background with Newton's name on the spine. Franklin, famous for his experiments on electricity, had more justification than most to link himself intellectually with Newton. The lieutenant governor of New York, a friend of Franklin and also a scientist of sorts, commemorated his purported explanation of the cause of gravity with a portrait of himself with his own book prominently at hand and a copy of Newton's *Principia* on a shelf behind him. The president of Yale College made no pretense to being a scientist, yet he too had a copy of the *Principia* included in his portrait, plus an emblem with comets moving in long ellipses. Newton's science had become an icon in art.

Figure 9.2: Benjamin Franklin with bust of Newton. Engraving by Edward Savage, 1793. National Portrait Gallery, Smithsonian Institution, Washington, D.C. (Library of Congress)

A particularly strong and extensive link to Newton exists in the writings of the Scottish economist Adam Smith (1723–1790). At the beginning of his youthful essay on *The Principles which lead and direct Philosophical Enquiries; illustrated by the History of Astronomy*, Smith expressed his belief that scientific hypotheses were but arbitrary creations. But then Smith moved forward in his history and studied the superior genius and sagacity of Sir Isaac Newton. Smith praised Newton's joining together of the movements of the planets by the universality of gravity into a philosophical system with a firmness and solidity no other system possessed. It was the greatest and most admirable discovery and improvement ever made by man. Furthermore, Newton's philosophical system constituted an immense chain of the most important and sublime truths all closely connected together.

In his 1776 *Wealth of Nations*, Smith presented real connecting principles of economics. Emulating Newton's strategy, Smith first listed phenomena; next he obtained general laws, ostensibly by induction; and then he deduced from the general laws both the listed phenomena and further phenomena. Newton's and Smith's answers to the question of the ultimate nature of their general principles are also similar. Unable to determine whether the propensity to exchange one thing for another was an original principle of human nature or a necessary consequence of the faculties of reason and speech, Smith argued that it was enough to know that the principle was universal. Furthermore, Smith employed Newton's principle of gravity both metaphorically and descriptively. The natural price was the central price to which the prices of all commodities were continually gravitating.

French economists also appreciated Newton. From the belief that government policy should not interfere with the operation of natural laws of economics came the phrase *laissez-faire, laissez-passer* (allow to do, let things pass).

Newton's influence also spread to the arts, especially to music. Ancient Greeks had sought a mathematical foundation for music, and music was part of the quadrivium taught in medieval universities, along with astronomy, arithmetic, and geometry. The French composer and musical theorist Jean-Philippe Rameau (1683–1764), who composed several operas for which Voltaire supplied the librettos, became known as the Newton of harmony. Rameau believed that music was a science that should have definite rules, that the rules should be drawn from an evident principle, and that the principle could not be known without the aid of mathematics.

Literature, too, was affected by Newton's scientific achievement. Voltaire lamented, as wittily as ever, that writing was hardly any longer the fashion in Paris, where everyone now worked at geometry and physics; it was as if Germans had conquered the country. Physics threatened to crush the flowers of poetry and to decompose the rainbow, warned the English Romantic poet Percy Bysshe Shelly (1792–1822). Physics threatened to deaden nature with a single vision, cautioned the poet and painter William Blake (1757–1827). Poets were uncomfortable with the excessive rationalism and rigid determinism of nature's laws. They preferred to speak to the heart and to the imagination. "O for a life of sensations rather than of thoughts," cried out the English Romantic John Keats (1795–1821) in a 1817 letter to a friend. "Bathe in the waters of life," urged Blake.

Still, even Romantics rebelling against Newtonianism recognized Newton's great achievement, and they expanded the concept of genius, previously reserved for artists and poets, to include Newton. The German dramatist and poet Friedrich von Schiller wrote that man had to be an animal before he knew that he was a spirit, he had to crawl in the dust before he ventured on the Newtonian flight through the universe.

CONCLUSION

If further explanation or justification is required for Thomas Jefferson's veneration of Newton as one of the greatest men who ever lived, it can be found in the pages of the French *Encyclopédie.* This analytical dictionary of the sciences, arts and trades, published between 1751 and 1772, was intended to unify and popularized the achievements of the new science. Jean D'Alembert (1717–1783), a French mathematician who contributed importantly to the advance of Newtonian science, and was also coeditor of the *Encyclopédie,* wrote:

> The true system of the world has been recognized ... natural philosophy has been revolutionized ... the discovery and application of a new method of philosophizing, the kind of enthusiasm which accompanies discoveries, a certain exaltation of ideas which the spectacle of the universe produces in us; all these causes have brought about a lively fermentation of minds. Spreading throughout nature in all directions, this fermentation has swept everything before it which stood in its way with a sort of violence, like a river which has burst its dams ... thus, from the principles of the secular sciences to the foundations of religious revelation, from metaphysics to matters of taste, from music to morals, from the scholastic disputes of theologians to matters of commerce, from natural laws to the arbitrary laws of nations ... everything has been discussed, analyzed, or at least mentioned. The fruit or sequel of this general effervescence of minds has been to cast new light on some matters and new shadows on others, just as the ebb and flow of the tides leaves some things on the shore and washes others away. (Buchdahl, *Image of Newton,* pps. 62–63)

Newton and his followers in many fields of human endeavor washed away more ancient thought and superstition than ever before in the tides of human affairs. They also washed in more new and revolutionary thinking. The new astronomy became intertwined with human values, culture, religion, and society. Its impact on Western civilization has been incredibly far reaching and profound, as wide as the ocean and as deep as the sea.

RECOMMENDED READING

Buchdahl, Gerd. *The Image of Newton and Locke in the Age of Reason* (London: Sheed and Ward, 1961).

Feingold, Mordechai. *The Newtonian Moment: Isaac Newton and the Making of Modern Culture* (New York and Oxford: The New York Public Library/Oxford University Press, 2004).

Hetherington, Norriss S. "Isaac Newton and Adam Smith: Intellectual Links between Natural Science and Economics," in Paul Theerman and Adele F. Seeff, *Action and Reaction: Proceedings of a Symposium to Commemorate the Tercentenary of Newton's* Principia (Newark: University of Delaware Press, 1993), pp. 277–291.

Keats, John. *Selected Letters.* (New York: Oxford University Press, 2002).

Newton, Isaac. *Mathematical Principles of Natural Philosophy and His System of the World By Sir Isaac Newton;* translated into English by Andrew Motte in 1729. The translation revised and supplied with an historical and explanatory appendix, by Florian Cajori (Berkeley: University of California Press, 1946).

Westfall, Richard S. *Never at Rest: A Biography of Isaac Newton* (Cambridge: Cambridge University Press, 1980).

———. *The Life of Isaac Newton* (Cambridge: Cambridge University Press, 1993).

WEB SITES

Feingold, Mordechai. *The Newtonian Moment: Science and the Making of Modern Culture*: http://www.nypl.org/research/newton.

———. *Isaac Newton and Thomas Jefferson*: http://www.monticello.org/streaming/speakers/newton.html.

The Newton Project: http://www.newtonproject.sussex.ac.uk/prism.php?id=1.

10

Astronomy and Religion

The creation, composition, and construction of the universe is the affair of astronomers, while humanity's situation and significance in the universe is a theme for theologians. When revolution in either of these branches of understanding calls for corresponding change in the other, conflict is possible. The interrelationship between astronomy and religion need not be antagonistic and destructive, however; it can be supportive, enhancing, even transformational.

Helpful interaction can ensue when advance in either intellectual realm encourages more intelligent introspection in the other, especially if the party under pressure to revise its tenets is accommodating to change. Both Saint Augustine (354–430) and Saint Thomas Aquinas (ca. 1225–1274), while insisting that the truth of Scripture is inviolable, also acknowledged that alternative interpretations of Scripture are possible and warned against rigid adherence to any particular interpretation. A hasty choice could prove detrimental to faith, were science later to prove that choice untenable.

Hostile interaction between astronomers and theologians can occur when either assembly of adherents makes dogmatic pronouncements in an area of theories or belief pertaining to the other. During the course of hearings in the New York State legislature in 1868 in connection with the enabling legislation for Cornell University, clergymen warned against the atheism of the proposed university with its emphasis on science. Andrew Dickson White (1832-1918), Cornell's first president, refused to stretch or cut science to fit revealed religion, and the dogmatic opposition he encountered encouraged and reinforced his belief in the inevitability of hostility between laws of science and articles of faith. Later, while United States

Minister to Russia from 1892 to 1894, White watched workers chip away at the ice barrier across the River Neva binding together the piers and the old fortress of the czars. He likened the ice to outworn creeds and noxious dogmas attaching the modern world to medieval conceptions of Christianity, and he wished that both ice and dogma might be swept away by floods, the former by water and the latter by increased knowledge and new thought. In all of modern history, White charged in his 1896 book *History of the Warfare of Science with Theology in Christendom*, interference with science in the supposed interest of religion had resulted in the direst evils both to religion and to science.

EARLY INTERACTION BETWEEN ASTRONOMY AND RELIGION

The Greek philosopher Anaxagoras reputedly was a victim of conflict between astronomy and religion. With Anaxagoras, Greek astronomy had moved away from astrological superstition, magical powers, and myth toward a more rational spirit and a universe with unchanging ways ascertainable by human reason. Anaxagoras's new theory of universal order collided with popular faith, the belief that gods ruled the celestial phenomena, and he was expelled from Athens.

Impiety, however, may have been an incidental charge. The indictment also included an accusation of corresponding with agents of Persia. Furthermore, enemies of the Athenian statesman Pericles (ca. 495–429 BC) may have sought to damage him by attacking his friend Anaxagoras. The Roman writer Diogenes Laërtius (ca. third century AD) in his book *Lives and Opinions of Eminent Philosophers* wrote:

> The Sun's a molten mass,
> Quoth Anaxagoras;
> This is his crime, his life must pay the price.
> Pericles from that fate
> Rescued his friend too late;
> His spirit crushed, by his own hand he dies.

The Romans celebrated suicide as a noble death, and Diogenes may have attributed such an honorable end to Anaxagoras more on the basis of convention than evidence. In any case, the rise of a new scientific attitude and mode of thought may well have accelerated the downfall of traditional religious and political beliefs, and also helped shape their replacements.

For Greek philosophers, astronomy was a religious undertaking. Plato's conception of cosmic order was permeated with ethical overtones and moral significance. Ptolemy, too, cultivated cosmology particularly with respect to divine and heavenly things. He wrote that he knew he was a mortal and a creature of a day, but when he searched out the massed wheeling circles of the stars, his feet no longer touched the Earth. Side by side with Zeus himself, Ptolemy took his fill of ambrosia, the food of the gods. Contemplation of the heavens was for Ptolemy a religious experience.

Similarly for Christians, the Heavens declared the glory of God (Psalms 19:1). Study of the cosmos, the perfect expression of divine creativity and providence, was a way to know God. Astronomy came to be regarded as a handmaiden, subservient to theology, pursued not for its own sake but for its usefulness in the interpretation of Holy Scripture.

Astronomical knowledge was not valuable enough, however, to prevent the Roman emperor Justinian (ca. 482–565), a Christian, from closing the Academy at Athens in 529 and forbidding pagans to teach. Teachers were being judged not for their intellect, but for their faith. Not all was lost, though. Of the philosophers forced to leave Athens for Persia, some returned in a few years, and the Academy may have continued in some less public form.

Competition from the church, which now offered a rival and more generously funded profession for bright young men, further impeded the pursuit of astronomical studies. Little patronage had been conferred on astronomy by ancient societies, with the notable exception of the Museum at Alexandria, and astronomy's precarious social position did not improve under early Christianity.

MEDIEVAL ISLAMIC ASTRONOMY AND RELIGION

In contrast to the West, where astronomy at least sometimes was accepted as useful in the interpretation of Holy Scripture, in Islamic countries astronomy was often perceived as dangerous to the faith. Consequently, the study of astronomy was not widely incorporated into Islamic educational institutions. Indeed, some Islamic philosophers even debated whether Islamic law prohibited the study of science.

When astronomy was taught, Islamic traditionalists criticized the establishment of schools for heretics and the teaching of magic. The Istanbul Observatory was torn down after plague, defeats of Turkish armies, and the deaths of several important persons were attributed to astronomers' attempts to pry into the secrets of nature. Yet astronomical studies persisted within Moslem culture and civilization, sometimes under individual rulers interested in astrology.

Islam was not a centralized religion to the extent that Western Christendom was, and Islam thus possessed a potential for greater intellectual freedom and diversity. On the other hand, church and state were not separate in Islamic countries, as they increasingly were in the West. This fact may help explain why science eventually developed so much further in the West.

MEDIEVAL CHRISTIAN ASTRONOMY AND RELIGION

Knowledge of Greek science almost disappeared from Western Europe between 500 and 1100, before it was recovered in translations of the works of Aristotle and Plato and in Arabic treatises and commentaries on Aristotle. After transmission from Islam to the West, Aristotelian astronomy fused with Christian theology into Scholasticism. In this form, Aristotle's astronomy permeated thought in Western Europe between roughly 1200 and 1500, especially in universities.

The Aristotelian position that there can be established necessary astronomical principles had the potential to result in truths necessary to astronomy but contradictory to dogmas of the Christian faith. For example, the Aristotelian assertion that the world had no beginning and no end, and thus was indestructible, seemingly conflicted with the Christian certainty of a Day of Judgment.

In 1270, the Bishop of Paris condemned several propositions derived from the teachings of Aristotle, including the eternity of the world and the necessary control of terrestrial events by celestial bodies. Seven years later, the bishop was directed by Pope John XXI to investigate intellectual controversies at the university. Within three weeks, the bishop condemned 219 propositions, including the claim that God could not make other worlds. Excommunication was the penalty for holding even one of the damned errors.

Although intended to contain and control scientific inquiry, the condemnation helped free astronomy from Aristotelian prejudices and modes of argument. Thus the Scientific Revolution may owe something to the condemnation of 1277. But probably not as much as is sometimes attributed to it, because not until the seventeenth century would astronomers repudiate Aristotelian science.

What the condemnation of 1277, with its emphasis on God's absolute power, undisputedly led to was the nominalist thesis (also called the instrumentalist thesis and the positivist thesis). Aristotelians had believed that necessary principles of astronomy and physics could be established. This position now was rejected. Astronomy was to be understood, instead, as a working hypothesis in agreement with observed phenomena. The truth of any particular hypothesis could not be insisted upon, because God could have made the world in a different manner, but with the same observational consequences. Astronomers might come to conclusions, but they could not insist that their conclusions were binding or limiting upon God's power to have created the world in a different way. Astronomical theories were tentative, not necessary, and thus could pose no challenge to religious authority.

While conceding the divine supremacy of Christian doctrine and acceding to religious authority, the nominalist thesis also freed science from religious authority. It was a convenient stance in a time when religious matters were taken seriously and heretical astronomical thoughts could place their adherents in serious danger from powerful ecclesiastical authorities. In the new intellectual climate, imaginative and ingenious discussions flourished.

Echoing and Mirroring the Divine Universe

Saint Augustine believed that the essence of beauty resided in resemblance to the divine universe. Seemingly discordant elements are brought into ultimate harmony under a single overarching geometrical order.

Beautiful music is composed of simple arithmetic ratios between musical notes. The result echoes the divine universe, which consists of similar ratios. The medieval motet, an unaccompanied choral composition with different texts (sometimes in different languages) sung simultaneously over a Gregorian chant fragment, originated in the thirteenth century. It was sung as part of church services, for the greater glory of God and for the enlightenment, instruction, and diversion of man. The music's intricate polyphonic form, in which every little detail has its place and meaning, supposedly echoed the superbly organized divine universe. One of the greatest expressions of this musical genre is found in the eighteenth-century fugues of Johann Sebastian Bach (1685–1750), Capellmeister and Director Chori Musici in Leipzig.

Figure 10.1: God Designs the Universe with a Compass. Österreichishe Nationalbibliothek, Vienna, Latin MMS, MS 2554, fol. 1r. (Erich Lessing/Art Resource, NY)

Resemblance to the divine universe as the essence of beauty also applies to architecture, especially to Gothic cathedrals, constructed between the twelfth and sixteenth centuries. Dedicating the new choir of the abbey church of Saint-Denis in 1144, Abbot Sugar called it the embodiment of the mystical vision of harmony that divine reason had established throughout the cosmos. The geometrical regularity and harmony of Gothic cathedrals was, he believed, as literal a depiction of a spiritual ideal as could be built in stone. Architecture mirrored the divine universe.

Indeed, some medieval architects were so convinced of the power of geometrical regularity to stabilize structures that when the Milan cathedral showed signs in the 1390s of collapsing, one solution proposed was to *increase* its height, in order to form one section into a perfect square. Fortunately, instead of trying to bring the cathedral into closer harmony with the divine universe, more buttresses were added. Other cathedrals did collapse, including Worcester in 1175, Lincoln in 1240, Beauvais—the tallest Gothic cathedral ever built—in 1282, Ely in 1321, and Norwich in 1361.

THE NEW ASTRONOMY AND RELIGION

However imaginative and ingenious hypothetical scientific theories may have been, they were not the stuff of revolution. Not until the goal of "saving the appearances," as the nominalist endeavor has been called, was replaced with the quest to discover physical reality was Aristotelian science replaced with a new worldview. Confidence that the essential structure and operation of the cosmos is knowable was a prerequisite to the work of Copernicus, Galileo, Kepler, and Newton. Inevitably, some of their necessary astronomical theories would conflict with Christian dogma.

An unauthorized foreword to Copernicus's 1543 *De revolutionibus orbium coelestium* presented his heliocentric theory as a convenient mathematical fiction. Copernicus, however, believed that he was describing the real motions of the world. He anticipated criticism from people who believed otherwise, but not from the Catholic Church, in whose service he had long labored as a canon, and in whose service he had advised the papacy on calendar reform. Indeed, Copernicus dedicated his book to Pope Paul III in the hope that his labors would contribute to correction of the ecclesiastical calendar.

Citation of Scripture against the new astronomy was not long in coming. Even before Copernicus's book was published, the German theologian and church reformer Martin Luther (1483–1546) warned that the fool Copernicus wished to reverse the entire science of astronomy. Luther had only to note that sacred Scripture stated that Joshua had commanded the Sun and not the Earth to stand still (Joshua 10:13).

Literal adherence to the Bible was the foundation of the Protestant revolt against Catholic religious hegemony. Prior to the Counter-Reformation, the Catholic Church was more liberal in its interpretation of the Bible and more accepting of Copernican astronomy. The new astronomy was taught in some Catholic universities, and it was used for the new calendar promulgated by Pope Gregory XIII in 1582 (see Chapter 5, Calendars).

Astronomical Phenomena as Religious Metaphor

One of the proposals from the Council of Trent, held in the middle of the sixteenth century in response to the Protestant Reformation, was that the arts should communicate religious themes. Artists should reveal the divine in nature and thus make religious experiences more conceivable and accessible.

Depiction of St. Benedict's vision was a particularly difficult problem. He was said to have beheld a flood of light more brilliant than the Sun. Then the whole world united into a ray of light. And then the soul of Germanus, Bishop of Capua, was carried into heaven by angels in a ball of fire. Later reports placed Germanus's death at the precise time of St. Benedict's vision.

The German painter and architect Cosmas Damian Asam (1686–1739) employed astronomical phenomena as a metaphor for divinity and inspiration, setting St. Benedict's vision during a total solar eclipse. Asam's painting was so successful that eclipses subsequently were incorporated into the Bavarian Benedictine visual tradition. Progressive improvements in Asam's depictions of St. Benedict's vision suggest that Asam carefully observed a series of solar eclipses. He could do so because accurate predictions of eclipse paths across Europe, based on Newton's laws of gravity and motion, had become available in the eighteenth century.

GALILEO

The clash between Galileo and Catholic authorities was neither necessary nor inevitable. And it was worsened by an unfortunate concatenation of circumstances. Galileo complained that the struggle was initiated by Aristotelian philosophers in Italian universities attempting to bring the Church into battle on their side against him. They fabricated a shield for their fallacies out of religion and the authority of the Bible.

A few individual priests charged that the motion of the Earth was contrary to the Bible. Cardinal Cesare Baronius (1538–1607) cogently replied that the Bible teaches the way to go to heaven, not the way the heavens go.

Galileo cited Saint Augustine's warning that no scientific doctrine should ever be made an article of faith, lest some better-informed heretic might exploit misguided adherence to the doctrine to impugn the credibility of proper articles of faith. It was good advice for an era in which new telescopic observations were being made almost nightly.

Galileo also appealed to Saint Augustine's authority in support of the thesis that no contradiction can exist between the Bible and science when the Bible is interpreted correctly. It was very pious to say, and most prudent to affirm, that the holy Bible could never speak untruth. But for discussions of physical problems, Galileo insisted that it was better to begin from sense-experiences and necessary demonstrations, rather than from the authority of scriptural passages.

Galileo's position may sound sensible now, but as a passionate fighter against dogma based on authority, he was out of step with his time. The Counter-Reformation then demanded tight control over Church doctrine, the better to counter Protestants.

In 1616, Pope Paul V submitted the questions of the motion of the Earth and the stability of the Sun to the official qualifiers of disputed propositions. Galileo expected them to read the Bible metaphorically. Instead, they read it literally. They found both the motion of the Earth and the stability of the Sun false and absurd in philosophy. They did not rule on the truth of Copernican astronomy. But they did rule that the motion of the Earth was at least erroneous in the Catholic faith, and the stability of the Sun was formally heretical.

The qualifiers had exceeded their authority, because only the pope or a Church council could decree a formal heresy. Consequently, the pope ignored their finding. The Congregation of the Index, however, issued an edict forbidding reconciliation of Copernicanism with the Bible and assertion of literal truth for the forbidden propositions. One passage about scriptural interpretation and passages calling the Earth a star (implying that it moved like a planet) were ordered removed from Copernicus's *De revolutionibus*. Catholics could still discuss Copernican astronomy hypothetically, and little damage had been done to science.

At a meeting with Church officials, Galileo was instructed no longer to hold or defend the forbidden propositions: the motion of the Earth and the stability of the Sun. Had he resisted, the Commissary General of the Inquisition was prepared to order Galileo, in the presence of a notary and witnesses, not to hold, defend, or teach the propositions in any way, on pain of imprisonment. Galileo did not resist, but the Commissary General may have read his order anyway. It appears in the minutes of the meeting, unsigned and unwitnessed. Galileo may have been advised to ignore the unauthorized intervention. Subsequent rumors that Galileo had been compelled to abjure caused him to ask for an affidavit—which he received—stating that he was under no restriction other than the edict applying to all Catholics.

In 1623, a new pope was chosen. Urban VIII was an intellectual, admired his friend Galileo, granted him six audiences in 1624, and encouraged him to write about the Copernican theory as a mathematical hypothesis. The resulting book, Urban hoped, would demonstrate that the Church did not interfere with the pursuit of science, only with unauthorized interpretations of the Bible.

Galileo's *Dialogue Concerning the Two Chief World Systems, Ptolemaic & Copernican* was published in Florence in 1632 under a formal license from the Inquisition. Presenting Galileo's views with biting wit is Salviati, named after Galileo's friend Filipo Salviati, a sublime intellect who fed no more hungrily upon any pleasure than he did upon fine meditations. Presenting arguments made personally to Galileo by the pope is Simplicio, named after the sixth-century Aristotelian philosopher Simplicius, who wrote commentaries on Aristotle's physics and astronomy. Simplicio is neither as smart nor as well informed as Salviati, and in the *Dialogue* Galileo ridiculed and destroyed Aristotelian physics. The third participant, an intelligent layman initially neutral in the debate, is Sagredo, named after Galileo's friend Giovanfrancesco Sagredo, of noble extraction and trenchant wit.

Not surprisingly, Urban grew angry when he found his own thoughts attributed to the Aristotelian representative in the *Dialogue* who lost every argument. Nor was it prudent to anger Urban. One summer he had had all the birds in the papal gardens in Rome slaughtered because the noise they made distracted him from his work.

How could Galileo have been so imprudent, even self-destructive? His nature was such that he could not have done otherwise. His passionate fight employing biting sarcasm against Aristotelian dogma created enemies all his life.

Figure 10.2: Frontispiece, *Dialogo di Galileo Galilei*, 1632. The title *Dialogo* is followed by Galileo's name and references to his membership in the Accademia dei Lincei and his employment under the Grand Duke Ferdinand II of Tuscany, son of Cosimo II, Galileo's earlier patron, who had died in 1620. The Academy of the Lynxes, founded in 1603 in Rome by an Italian nobleman passionately interested in science, and since 1871 Italy's official academy of sciences, had chosen the sharp-eyed lynx, and also the eagle, as symbols emphasizing the importance of observation. Image courtesy of History of Science Collections, University of Oklahoma Libraries; Copyright the Board of Regents of the University of Oklahoma.

Also, the timing was unfortunate for Galileo. His book appeared during a period of heightened suspicion, and even paranoia. A Spanish cardinal had criticized Urban for interfering in a political struggle, and Urban had responded with a purge of pro-Spanish members of his administration, including the secretary who had secured permission for printing Galileo's *Dialogue*.

Galileo was called to Rome, where he was charged with contravening the (unsigned and unauthorized) order of the Inquisition not to hold, defend, or teach the Copernican propositions in any way. In response, Galileo produced his affidavit, signed and dated, stating that he was under no restriction other than the edict applying to all Catholics. Nonetheless, he was found guilty and compelled to abjure, curse, and detest his errors and heresies. Henceforth, even hypothetical discussion of Copernican astronomy was heresy for Catholics.

The Galileo fiasco has long been an embarrassment for the Catholic Church. In 1978, Pope John Paul II acknowledged that Galileo's theology was sounder than that of the judges who had condemned him. In 1981, John Paul set up a special committee to reexamine the Galileo case. And in 1992, acting on the findings of that eleven-year investigation, John Paul lifted the edict of the Inquisition against Galileo.

KEPLER AND NEWTON

Kepler had been convinced that the Creator used mathematical archetypes to design the universe, and this conviction drove Kepler's research and shaped his results. Kepler's laws were a sacred sermon, a veritable hymn to God the Creator. Kepler's laws showed how great was God's wisdom, power, and goodness.

Newton, too, was convinced that he was exploring and demonstrating God's wonders. Newton had found it unlikely that random chance could have been responsible for the formation of the solar system, and he stated in his *Principia* that the most beautiful system of the Sun, planets, and comets could only have proceeded from the counsel and dominion of an intelligent and powerful Being. Nothing better conveys the concept now called intelligent design, although current arguments are centered in biology and directed against the theory of evolution. Even when Newton realized that irregularities in planetary motions caused by disturbing influences of planets upon each other would increase until the entire system needed adjustment, he chose to construe the necessity of God's occasional reformation of the solar system not as a flaw in God's power but as further evidence of His existence.

Many of Newton's contemporaries also were convinced that he had demonstrated God's wonders. The Scottish poet James Thomson (1700–1748) asked in *To the Memory of Sir Isaac Newton,* published only three months after Newton's death:

> Shall the great soul of Newton quit this earth.
> To mingle with his stars; and every Muse,
> Astonish'd into silence, shun the weight
> Of honours due to his illustrious name?

Thomson answered himself in his poem *The Seasons*:

> Newton, pure Intelligence, whom God
> To Mortals lent, to trace his boundless Works
> From laws sublimely simple.

Newton was the first scientist honored with burial in Westminster Abbey. Above the reclining figure of Newton sculpted in marble, his left hand pointing to a scroll held by two winged boys and with a mathematical design on it, is a large celestial globe with signs of the zodiac and constellations, and the path of the comet of 1680. On top of the globe is the Greek goddess Urania, the muse of astronomy, leaning on a book.

In the mystery thriller *The Da Vinci Code* by Dan Brown, the protagonists are looking for the tomb of a knight a pope interred. Eventually they remember that the poet Alexander Pope had eulogized Newton with the couplet "Nature, and Nature's Laws lay hid in Night. / God said, 'Let Newton be!' and all was Light." It is Newton's tomb at Westminster they seek.

On his tomb Newton is praised as having vindicated by his philosophy the majesty of God mighty and good. The inscription (in Latin) at the bottom of the tomb reads, in English translation:

> Here is buried Isaac Newton, Knight, who by a strength of mind almost divine, and mathematical principles peculiarly his own, explored the course and figures of the planets, the paths of comets, the tides of the sea, the dissimilarities in rays of light, and what no other scholar has previously imagined, the properties of the colours thus produced. Diligent, sagacious and faithful, in his expositions of nature, antiquity and the holy Scriptures, he vindicated by his philosophy the majesty of God mighty and good, and expressed the simplicity of the Gospel in his manners. Mortals rejoice that there has existed such and so great an ornament of the human race.

Notwithstanding all the exuberant praise heaped on Newton for his demonstration of God's majesty, theological implications of Newton's astronomy were criticized by the German philosopher and mathematician Gottfried Wilhelm Leibniz (1646–1716). He sent his complaint to Caroline, Princess of Wales. Newton's friend Samuel Clarke (1675–1729), a British philosopher, answered the charge. His letter also was addressed to Caroline, and she forwarded it to Leibniz.

In the course of the debate, Leibniz wrote five letters and Clarke wrote five replies, all of which were published. Leibniz charged that Newtonian views were contributing to a decline of natural religion in England, because the implication that God occasionally intervened in the universe, much as a watchmaker might wind up and mend his work, derogated from His perfection. Clarke admitted that God had to intervene in the universe, but only because intervention was part of His plan. Indeed, the necessity of God's occasional reformation was, for Newtonians, proof of God's existence.

ASTROTHEOLOGY AND GALAXIES

Eighteenth-century belief in the orderliness of the universe made determination of that order an important theological, philosophical, and scientific endeavor for astrotheologians. William Whiston (1667–1752), Newton's successor in the Lucasian Chair of Mathematics at Cambridge University from 1702 to 1710, until he was charged with heresy and dismissed from the university, argued that the system of the stars, the work of the Creator, had a beautiful proportion, even if frail man were ignorant of the order.

William Derham (1657–1735), an ordained priest in the Church of England, a vicar, and royal chaplain, expressed a similar belief. In his 1715 book *Astrotheology; or, a Demonstration of the Being and Attributes of God, from a Survey of the Heavens*, Derham wrote that there was a great Parity and Congruity observable among all the works of the Creation, and they had a manifest harmony and great agreement with one another.

Newtonian gravitational theory practically demanded a continual miracle to prevent the Sun and the fixed stars from being pulled together. If the stars were moving in orbits around the center of a system, however, like the planets move around the Sun, the result could be a stable system rather than gravitational collapse. In 1718, the English astronomer Edmund Halley reported that three bright stars were no longer in the positions determined by ancient observations. The weight of tradition was so heavy, however, that even Halley continued to call the stars "fixt stars."

Eventually, Halley's discovery suggested to the self-taught English astronomer Thomas Wright (1711–1786) that the stars might be revolving around their center of gravity, and thus were prevented from falling into their center, just as the planets are prevented from falling into the Sun. In 1750, Wright proposed such a model for the Milky Way, a luminous band of light observed circling the heavens.

The German philosopher Immanuel Kant (1724–1804) was inspired by an incorrect summary of Wright's book and by the paradigm of the Newtonian solar system. Kant's explicitly expressed intent was to extend Newtonian philosophy. The subtitle of Kant's *Universal Natural History and Theory of the Heavens* was *An Essay on the Constitution and Mechanical Origin of the Whole Universe Treated According to Newton's Principles.*

Kant explained the Milky Way as a disk-shaped system seen from the Earth, which was located in the plane of the disk. The arrangement of the stars, thought Kant, might be similar to that of the planets. Furthermore, the Newtonian system provided by analogy a physical explanation for a disk structure. Kant reasoned that the same cause that gave the planets their centrifugal force and directed their orbits to a plane could also have given the power of revolving to the stars and brought their orbits into a plane. Thoroughly imbued with a belief in the order and beauty of God's work, Kant went on to suggest that nebulous patches of light in the Heavens are composed of stars and are other Milky Ways, or island universes (galaxies similar to and beyond the boundaries of our own galaxy).

The Milky Way

An observer located at the center of a thin stratum of stars will see the surrounding stars projected as an encircling ring. William Herschel (1738–1822), who had left Germany and made his living as a musician in England before becoming the world's preeminent amateur astronomer, reported that clusters of stars are arranged into strata, and that the Milky Way, an irregular band of light circling the celestial sphere, is the projection of a stratum of stars and groups of stars.

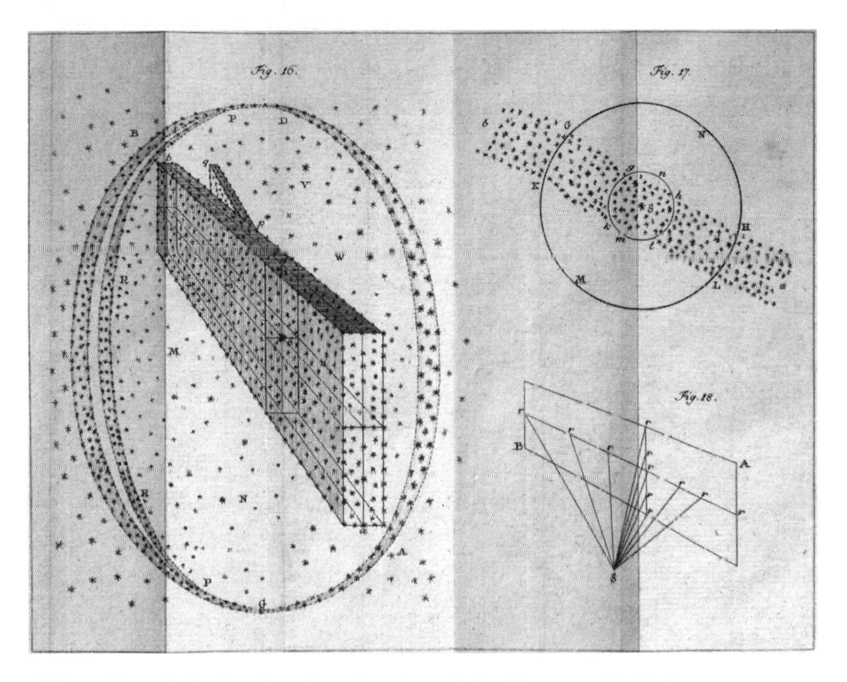

Figure 10.3: William Herschel, *On the Construction of the Heavens*, 1784. Image courtesy of History of Science Collections, University of Oklahoma Libraries; Copyright the Board of Regents of the University of Oklahoma.

CHANGING WORLDVIEWS

In the period between Whiston and Kant, a major change in worldview occurred. At the beginning of the eighteenth century, the stable structure of the universe, as well as its initial creation, was believed to be the work of God. As the century progressed, God's role diminished. By the end of the century, laws of nature and sequences of mechanical events explained evolution from an initial chaos to the then-observed cosmological structure. As Laplace told Napoleon (see Chapter 9, The Newtonian Revolution), he had no need for the hypothesis of a Creator. Astronomy and Christianity, formerly joined in Western thought, now were estranged. The divorce of God from the physical universe may well have been inevitable as the Newtonian revolution played out, however convinced were its founding fathers that they were exploring and demonstrating God's wonders.

The perceived significance of humans and their particular planet was significantly diminished early in the twentieth century, when the American astronomer Harlow Shapley (1885–1972) established a new understanding of the universe. He determined it to be ten or even a hundred times larger than previously believed. Furthermore, Shapley found the Sun far toward one edge of the galactic plane, not near the middle.

Shapley noted a historical progression from belief in a small universe, with humankind at its center, to a larger universe, with the Earth far from the center. The geometry of the universe had been transformed from anthropocentric (ancient man) to geocentric (Ptolemy) to heliocentric (Copernicus) to acentric. The psychological change was no less, Shapley insisted, from homocentric to acentric.

For all the revolutionary implications of his pioneering observations, Shapley still believed in a single, enormous, all-comprehending galaxy, rather than a plurality of stellar universes. A few years later, however, Edwin Hubble (1889–1953), who like Shapley did his revolutionary work at the Mount Wilson Observatory in Southern California, determined that the distance to the spiral Andromeda Nebula was roughly a million light years, several times more distant than Shapley's estimate of the outer limits of our own galaxy. On February 19, 1924, Hubble wrote to Shapley to inform him of the distance determination. On reading Hubble's letter, Shapley remarked to a colleague who happened to be in his office that this letter destroyed his universe. Henceforth, astronomers understood that the universe is composed of innumerable galaxies spread out in space farther than the largest telescope can see.

Hubble next showed that the universe is not static, as nearly everyone then believed, but is expanding. What he had made infinite in space, he would make finite in time. In each of these successive worldviews, the system of stars was larger than had been realized before. The likely significance of humans and their planet in the sidereal scheme decreased as the dimensions of the discernible stellar universe increased.

Human stature was also under threat from the growing realization that it is unlikely that the solar system is anywhere near the center of our galaxy, let alone near the center of the entire universe. The presumption of random distribution leaves Earth's inhabitants in all likelihood far removed from a privileged place in the middle of an immense cosmos.

For people clinging to an anthropocentric view of the universe, contact with extraterrestrials could prove psychologically devastating (see Chapter 7, Extraterrestrial Life and Science Fiction). Questions about the possible existence of intelligent life elsewhere in the universe and whether extraterrestrials would be evidence of God's omnipotence and benevolence are yet to be answered.

ROLE REVERSAL

For many people, science has replaced religion as the source to which they turn for inspiration, for direction, and for criteria of truth. Religion, and also economics, politics, sociology, and many other disciplines, now appeal to science—especially to astronomy and cosmology—for legitimacy and validation.

One appeal of religion to science occurred with the Catholic Church's claim to Georges Lemaître (1894–1996) and the expanding universe theory he championed. Lemaître offered a second chance for the Catholic Church to embrace and be embraced by a second Galileo.

Lemaître was a Belgian astrophysicist, a Catholic priest, and, from 1960 until his death, president of the Pontifical Academy of Sciences. He was well aware that any discussion of the beginning of the world, inherent in his conception of a big bang and subsequent expansion, would raise the issue of the creation, but he tried to avoid direct conflict between astronomy and religion. In the 1930s, he wrote that everyone who believed in a supreme being believed also that God was essentially hidden, that present physics provided a veil hiding the creation. In the 1950s, in response to a statement by the pope, Lemaître insisted that science and theology are separate fields and that his astronomical theory stood outside any metaphysical or religious question.

In contrast to Lemaître, the British astronomer Fred Hoyle (1915–2001) intentionally emphasized conflict between astronomy and religion. A great opportunity for Hoyle to further his antireligious polemic occurred in 1951, when Pope Pius XII stated that the modern big bang theory affirmed the notion of a transcendental creator. The pope also associated the rival steady state theory with atheism. Hoyle was a leading creator and proponent of the steady state theory, in which there was no room for a Creator.

Advisors to the pope quickly convinced him that linking astronomy and theology was helpful to neither, and the pope never again looked to astronomy for support of Christian dogma. But the one incautious utterance was enough of an opening for Hoyle. He seized upon the pope's linkage of big bang cosmology with Christian dogma, finding this yet another reason to reject big bang theory as an irrational process beyond the realm of science.

In politics, too, temptations arise to appeal to astronomical theories for validation or refutation. Another early proponent of an expanding universe, the Russian mathematician Alexander Friedmann (1888–1925), was hailed as an example of great Soviet science and, by implication, vindication of the Soviet political system (no matter that difficult conditions in revolutionary

Russia in the 1920s and Friedmann's early death from typhoid fever severely limited his scientific output).

During Stalin's rule, Soviet astronomers were expected to serve the party by providing an astronomy congruent with official party ideology. The expanding universe theory and relativistic cosmology, however, became politically suspect, as bourgeoisie idealism and capitalist myth. Furthermore, the pope's 1951 statement suggested that the big bang theory was a religious, rather than a scientific, view. The rival steady state theory proposed infinity in both space and time, and thus lacked the theistic implications of a finite time scale and creation of the world from nothing, which were so objectionable to Marxists. But steady state theory, too, was unacceptable to Marxists; continuous creation of matter out of nothing tasted too much of religion and idealism. In the difficult political climate from the 1930s through the 1950s, Soviet astronomers generally avoided the ideologically sensitive study of cosmological models.

INTELLIGENT DESIGN REDUX

The concept of intelligent design, banished by Laplace in the case of the solar system, has come up again for the universe as a whole. Contemplating the remarkable set of physical coincidences that seems to have been necessary for life as we know it to exist, philosophers have come up with an equally remarkable label for their thoughts: the anthropic cosmological principle.

In its weak form, the anthropic principle states that the universe must be such as to admit and sustain life. From Descartes' "I think, therefore I am," we proceed to "I am, therefore the nature of the universe permits me to be."

In its strong form, the anthropic principle states that the universe was created and fine-tuned so intelligent life could evolve in it. This is the philosophical equivalent in astronomy of intelligent design in biology.

The most remarkable appearance of divine design is the value of the mass density of the universe. Had the initial density been different by as little as one part in ten to the sixtieth power, all matter would long ago have been crushed beyond recognition in the contraction of a big crunch, or torn apart beyond recognition in the expansion of a big chill. There would have been no time for planets to form and living creatures to evolve; there would be no intelligent life to contemplate the fact that the density is precisely what is needed to escape from oblivion.

Nor is the density of the universe the only remarkable numerical parameter arguing for divine intervention. The creation of stars, planets, and life would also be impossible were any single one of the values of various physical constants, including the strengths of fundamental forces such as gravity and electromagnetism, and the masses and charges of subatomic particles, the slightest bit different from what they are. Even Hoyle, notorious in the 1950s for his antireligious views, came to believe in the 1980s that it seemed that a superintellect had fiddled with astronomy, physics, chemistry, and biology. Blind forces seemed to Hoyle inadequate to explain the numbers calculated from observed facts.

New theories of particle physics, however, soon pointed to an explanation of the density of the universe as a natural and inevitable consequence, rather than the result of divine intervention and intelligent design. In the standard big bang model, density is an arbitrarily set parameter. In the new inflationary universe theory, however, a brief burst of exponential expansion in the first minuscule fraction of a second of the universe's evolution automatically drives the density, whatever its initial value, incredibly close to the critical density, at which neither a big crunch nor a big chill will occur.

Astronomical theories live or die by the congruence of their predictions with observation, in stark contrast to intelligent design, whose only test is its literal congruence with Scripture. Our galaxy is a gravitational clumping in a generally uniform cosmos. How did the tiny primordial nonuniformities of matter necessary to begin the process of gravitational clumping come about? Inflation could have stretched quantum mechanical fluctuations enormously, and the resulting wrinkles in space-time could have encouraged large-scale cosmic structures, such as a galaxy or even a cluster of galaxies, to coalesce. In 1982, NASA approved funding for the Cosmic Background Explorer Satellite and a search in the cosmic background radiation for tiny differences from uniformity predicted by the inflationary theory. Eventually, the experiment involved over a thousand scientists and cost more than $160 million. In 1992, astronomers reported finding 15-billion-year-old wrinkles in the background radiation, of the size predicted for seeds out of which our complex universe could have grown.

Astronomers are busy seeking plausible physical hypotheses and mechanisms to account for the remarkable structure of the universe otherwise attributable to God. Laplace's earlier success for the solar system augurs well for the current effort to encompass the entire universe. Repeatedly throughout history, astronomical complexities so remarkable that initially they were attributed to God eventually have been explained by plausible physical mechanisms.

CONCLUSION

The current battle over intelligent design is between questioning scientists seeking and testing naturalistic explanations of phenomena and a few opponents invoking religion to entangle unwitting allies in an effort to replace inquiry and reason with superstition. History shows that linking religious belief to any temporary stage of advancing scientific knowledge is likely to prove hazardous to religion. It remains to be seen how many individuals and institutions oblivious to history, and thus doomed to repeat it, will once again enlist on the losing side.

Cardinal Baronius demonstrated his understanding of a rational and rewarding relationship possible between science and religion when he said that the Bible teaches the way to go to heaven, not the way the heavens go. This aphorism has remained topical for centuries because it is witty and because it respects both science and religion.

RECOMMENDED READING

Alexander, H. G. *The Leibniz-Clarke Correspondence. Together with Extracts from Newton's* Principia *and* Opticks, *Edited with Introduction and Notes* (Manchester, UK: Manchester University Press, 1956).

Drake, Stillman. *Discoveries and Opinions of Galileo: Translated with an Introduction and Notes* (New York: Doubleday, 1957).

Ferngren, Gary B., ed. *The History of Science and Religion in the Western Tradition: An Encyclopedia* (New York: Garland Publishing, Inc., 2000).

Grant, Edward. *Science and Religion, 400 B.C. to A.D. 1550: From Aristotle to Copernicus* (Westport, CT: Greenwood Press, 2004).

Hetherington, Norriss S. ed., *Cosmology: Historical, Literary, Philosophical, Religious, and Scientific Perspectives* (New York: Garland Publishing, 1993).

Kragh, Helge. *Cosmology and Controversy: The Historical Development of Two Theories of the Universe* (Princeton, NJ: Princeton University Press, 1996).

Lindberg, David C., and Ronald L. Numbers, eds. *God and Nature: Historical Essays on the Encounter between Christianity and Science* (Berkeley: University of California Press, 1986).

WEB SITE

Cosmic Journey: A History of Scientific Cosmology: http://www.aip.org/history/cosmology/ideas/island.htm.

Epilogue

Stonehenge proclaims the existence of civilization in ancient Britain, pyramids memorialize Mayan culture in Central America, and Gothic cathedrals commemorate the Middle Ages in Europe. Modern Western culture may be similarly immortalized in its astronomical remnants, relics, and residue. Earthbound observatories boast telescope mirrors weighing tens of tons and domes weighing hundreds of tons, their movements triggered many times a second and completed in milliseconds, while computer-controlled pneumatic pistons simultaneously counter changes in gravitational stress on the mirrors as they assume different positions. Astronomical research outposts on the Moon, on Mars, and on large asteroids traversing the solar system are also likely to impress archaeoastronomers in some future era.

Astronomy reflects much of the cultures in which it has been embedded, and has also affected and been shaped by its cultural surroundings. The goals of astronomers, their theories, their instruments and techniques, their training, their places of work, and their sources of patronage have all undergone changes as revolutionary as anything else in the history of human cultures and civilizations.

Prior to the Scientific Revolution of the seventeenth century, the primary problem for astronomers in the Western world had been to discover the true system of uniform circular motions believed to underlie the observed and seemingly irregular motions of the planets, the Sun, and the Moon. Astronomers observed and recorded a few planetary, solar, and lunar positions, attempted to fit geometrical models to the observations, and constructed tables of positions. In addition to the effort for its own sake, there were practical offshoots, including personal horoscopes, more general warnings of man-made and natural catastrophes, and calendars foretelling times of religious celebrations and the agricultural seasons. Solar and stellar navigation became important only later; fourteenth- and fifteenth-century explorers still found their way in sight of land. The rest of the universe scarcely existed for ancient and medieval astronomers, other than as the limiting outer sphere of the stars. Nor did astronomers concern themselves with the physical composition of the universe.

The Copernican revolution radically redefined the astronomical agenda. After new observations refuted the ancient assumption of circular motion, the physical nature and cause of orbital motion—previously outside the province of astronomers—was of crucial importance. Interest in the physical composition of the heavens increased as well, once the Earth was displaced from the center of the universe, and especially after Galileo's revolutionary new telescopic observations. Henceforth, the discovery of hitherto unknown planets, moons, asteroids, comets, and nebulae, as well as examination of their more prominent features, was a standard activity of astronomers using telescopes. Amateur astronomers wealthy enough to procure relatively large telescopes could excel in this new milieu.

After some resistance, astronomers accepted the telescope as the primary instrument of astronomical measurement. Increasingly greater expenditures necessary for larger telescopes, for their operation, and for analyses of floods of new data all speeded transition from individual observers to observatories enjoying more ample government patronage. Tycho Brahe's positional measurements at the end of the sixteenth century were the last great pre-telescopic astronomical achievement. It was funded by Tycho's inherited wealth and by royal patronage bestowed in exchange for fame and glory. The Paris Observatory, founded as part of the new French Academy of Sciences in 1666, was generously funded by the state, with the hope that its work would contribute to the expansion of France's maritime power and international trade. England's Greenwich Observatory began operation a decade later under royal warrant. Its mission in the new age of exploration was to correct tables of planetary motions and stellar positions, so that longitudes could be determined and the art of navigation perfected.

Isaac Newton treated celestial motions as problems in mechanics governed by the same laws that determine terrestrial motions, and he recast the primary problem for astronomers from measuring positions to calculating orbits. NASA scientists now use computers to calculate trajectories for their spacecraft. Astronomers would be willing, for additional government funding, to calculate whether passing asteroids will safely miss the Earth.

The business of astronomy has at times been a family profession. Gian Domenico Cassini (1625–1712), recruited from Italy to set up the Paris Observatory in 1669, was succeeded by a son, a grandson, and a great-grandson, before the French Revolution drove the family from the observatory in 1793. Friedrich von Struve (1793–1864) also founded an astronomical dynasty spanning decades. He helped Czar Nicholas I (1796–1855) chart his vast empire and was the director of the Pulkovo Observatory, opened in 1839 outside St. Petersburg. A son succeeded Struve at Pulkovo; a grandson directed the Königsberg Observatory in Prussia; and a great-grandson, after fleeing revolution in Russia in 1921, directed first the Yerkes Observatory of the University of Chicago and later the astronomy department of the University of California at Berkeley.

In the middle of the nineteenth century, the Pulkovo Observatory shared the honor of possessing the world's largest refracting telescope, which bent to a focus light passing through curved glass lenses. As late as the end of the eighteenth century, glass of the quality necessary for optical instruments could be cast only in small pieces, up to two or three inches in diameter, and the English duty on manufacturing limited production and made further experimentation costly. Progress occurred on the continent, but even here the largest lens achieved by 1824 was a 9.5-inch disc. In 1847, the 15-inch refractors at Pulkovo and at the Harvard College Observatory were the largest in the world.

The British industrial revolution facilitated construction of the first large reflecting telescope, using metal mirrors to reflect light to a focus. William Herschel (1738–1822) cast a 48-inch metal mirror in 1789, and in 1845 in Ireland, William Parsons (1800–1867), the third Earl of Rosse, completed his leviathan, with a metal mirror six feet in diameter. Large reflectors with their tremendous light-gathering power yielded remarkable observations of distant stellar conglomerations, but difficulty in aiming tons of metal and rapid tarnishing of the mirrors rendered early reflectors unsuitable for observatories, which could not consistently harness the instruments' raw power.

New at the beginning of the twentieth century was the sixty-inch reflecting telescope built and installed at the new Mount Wilson Observatory near Los Angeles in 1908. The telescope had a reflective silver coating on a glass disc ground to bring incoming light to a focus, and a mounting system and drive capable of keeping the multiton instrument fixed on a celestial object while the Earth turned beneath it. The mountain observatory, funded by Andrew Carnegie's philanthropic Carnegie Institution, was one of the first located above most of the Earth's obscuring atmosphere. Its reflecting telescope, specifically designed for photographic work, completed the revolution in astronomical practice from the tedious drawing by hand of features seen through telescopes to virtually instant, and more objective, photographs.

With professional astronomers already employed in traditional research projects, it was largely left to amateur astronomers to pioneer the application of photography to astronomy. The American physician John Draper (1811–1882) took the first known photograph of a celestial object, the Moon, in the 1840s. In 1851, a new process using plates exposed in a wet condition made possible a few photographs of the brightest stars. Not until the introduction of more sensitive dry plates after 1878, however, did photography become common in astronomical studies. Draper's son Henry (1837–1882), also a doctor, in 1880 took the first photograph of a nebula.

The other major new astronomical tool of the nineteenth century was spectroscopy, which led to the new science of astrophysics. When attached to telescopes, prisms splitting light into different colors of the spectrum opened to investigation the physical and chemical nature of stars, because each chemical element produces its own pattern of spectral lines. The

English amateur astronomer William Huggins (1824–1910), who at age thirty sold the family business and built a small private observatory, went on to identify several elements in spectra of stars and nebulae. He also measured motions of stars, as revealed by slight shifts of their spectral lines.

Early in the twentieth century, Vesto M. Slipher (1875–1969) at the Lowell Observatory in Arizona was the first to measure shifts in spectra of faint spiral nebulae, whose receding motions revealed the expansion of the universe. An extended photographic exposure over three nights was required to capture enough light for the measurement.

Astronomical entrepreneurship saw the construction of new, larger instruments, and a shift of the center of spectroscopic research from England to the United States. Charles Yerkes (1837–1905), who controlled much of Chicago's street railway system, and James Lick (1796–1876), whose real estate speculations made him the wealthiest man in California, put up the funds for their eponymous observatories, which came under the direction, respectively, of the University of Chicago and the University of California. Percival Lowell (1855–1916), a wealthy Boston businessman, founded and directed his own observatory in Arizona. All three observatories were far removed from cities, and the latter two were perched on mountain peaks.

A scientific education was necessary for professional astronomers by the end of the nineteenth century, when astrophysics became important and the interests of professional and amateur astronomers diverged. As late as the 1870s and 1880s, the self-educated American astronomer Edward Emerson Barnard (1857–1923), an observaholic possessed of indefatigable energy and sharp eyes, could earn a place for himself at a major observatory with his visual observations of planetary details and his discovery of comets and the fifth satellite of Jupiter, but he was an exception and an anachronism.

A project begun in 1886 at the Harvard College Observatory to obtain photographs and catalog stellar spectra furthered another social shift in astronomy. The observatory employed women, albeit for lower wages than men would have received. But at least a door was opened into a formerly male-only profession.

Annie Jump Cannon (1863–1941) was largely responsible for the *Henry Draper Catalogue*, published between 1918 and 1924. It reported spectral type and magnitude for some 225,000 stars. She also rearranged the previous order of spectra into one with progressive changes in the appearance of the spectral lines. Although Cannon had developed her spectral sequence without any theory in mind, astronomers quickly realized that changes in the strength of hydrogen lines indicated decreasing surface temperature. Spectral class became even more meaningful when it was related to observed luminosity, which is an indication of the intensity of light and thus the distance of the source from the observer.

The production of stellar energy, and with it the constitution of stars, became better understood after the discovery of radioactivity. As late as

1920, however, the English astrophysicist Arthur Eddington (1882–1944) complained that the inertia of tradition was delaying acceptance of the most likely source of stellar energy: the fusion of hydrogen and helium. Research during World War II produced a deeper understanding of nuclear physics and more powerful computational techniques. The practice of astrophysics moved from observatories to scientific laboratories, and then to giant computers running simulations.

Cosmology, the study of the structure and the evolution of the universe, only belatedly insinuated itself into modern mainstream astronomy. And only later, yet, did it connect with Einstein's relativity theory. Astronomers, especially in the United States, possessed little mathematical training, and were largely content to produce observations, while leaving theory to theoreticians. In England, in contrast, interest in relativity theory relatively flourished. The work was primarily mathematical, however, with little observational input. Different national fashions in astronomy were encouraged and enforced by disparate access to large telescopes.

During the 1970s and 1980s, it was realized that important features of the universe could be explained as natural and inevitable consequences of new theories of elementary particle physics, and this science now increasingly drives advances in cosmology. Also, particle physicists, having neared the limits of current particle accelerators and of public funding for ever-larger instruments, now turn to astronomy for information regarding the behavior of matter under extreme conditions, such as prevailed in the early universe.

The greatest change in astronomy during the twentieth century in understanding the universe and also in the backgrounds of astronomers and the nature of their activities followed from observations not involving visual light. In the 1930s, the American radio engineer Karl Jansky (1905–1950) and the radio amateur Grote Reber (1911–2002) pioneered detection of radio emissions from celestial phenomena. Radar research during World War II advanced radio astronomy, especially in England.

X-ray astronomy took off after World War II, first aboard captured German V-2 rockets carrying detectors above the Earth's absorbing atmosphere. The subsequent cold war and the Soviet success in first placing a satellite in orbit around the Earth necessitated a grand American space program. The National Aeronautics and Space Administration funded a rocket survey program and then satellites to detect X-rays. Most of the new X-ray astronomers came over from experimental physics with expertise in designing and building instruments to detect high-energy particles, and their discoveries followed from technological innovations.

Gamma rays and infrared light and ultraviolet light have revealed more novelties in space. Unlike the relatively quiescent universe observed by earthbound astronomers using visual observations, the universe newly revealed to engineers and physicists observing from satellites at other wavelengths is violently energetic.

Since the 1970s, physicists have worked on string theory, in which the basic constituents of the universe are tiny wriggling strings rather than particles. The hope is that string theory will explain all the laws of physics and all the forces of nature in a single equation. Astronomers, however, have found no experimental evidence for string theory, nor have they thought up any prediction of observations that could refute it. Nonetheless, the theory is too beautiful to be allowed to die. If some of the original submicroscopic strings were stretched as large as galaxies during the brief inflationary spurt of the early universe, they might now be snapping under enormous tension, and the resulting ripples in space-time might be detectable by the new Laser Interferometer Gravitational Wave Observatory, which began observations in 2005. NASA bore its $365 million cost.

New kinds of particles might be needed to answer the question of missing matter. The estimated gravitational pull of visible matter is not sufficient to hold fast-moving galaxies together. Furthermore, the outer stars of galaxies are orbiting so fast that they should fly off, if nothing binds them other than the gravitational pull of the visible stars. Astronomers see only about 10 percent of the mass in the universe. Some of the invisible universe may consist of stars that emit very little light, such as brown dwarfs or primordial black holes. Weakly interacting massive particles, predicted by theoretical physicists, might also constitute some of the missing matter, but they have not been detected in the laboratory. Other theories point to topological defects in space-time, cosmic wrinkles left over from the early universe.

In addition to dark matter, there seems to be dark energy as well. It may be causing the expansion of the universe to accelerate, perhaps to the extent that it will rip apart galaxies and even matter itself in the far-distant future.

Interrelationships between astronomy and culture are more readily apparent in earlier eras than in our own time. Clashes and congruencies of human values with scientific theories in previous times are not hidden among our modern predispositions and predilections, which we perceive as normal rather than as the cultural constructs they are. Yet the goals of astronomy, its theories, its instruments and techniques, the training and the places of employment of its members, and sources of patronage are all interlinked with other aspects of modern civilization and culture. Astronomy continues to reflect much of the culture in which it is embedded, both shaped by and shaping its cultural surroundings.

Astronomical allusions, too, continue to link science with the humanities. Abraham Lincoln said that to add glory to the name of George Washington was as impossible as it was to add brightness to the Sun; none should attempt it, but in solemn awe merely pronounce the name and leave it shining on. Another, lesser politician, Jimmie Davis, governor of Louisiana from 1944 to 1948 and 1960 to 1964, and also a singer and songwriter, wrote the song "You Are My Sunshine." It became an official state song of Louisiana, received a Grammy Hall of Fame award, was named one of the songs of the century, and is ranked one of the greatest songs in country music. In it the singer sentimentally equates his beloved with the sunshine that makes him happy when skies are gray. In the 1960s, the folk-singing group Peter, Paul, and Mary popularized "Lemon Tree," a song in which the beloved's smile caused the Sun to rise in the sky, and her departure took away the Sun. The metaphors in these songs may fall short poetically of Shakespeare's likening of Juliet to the Sun ("But, soft! What light through yonder window breaks? It is the east, and Juliet is the Sun"). They are, nonetheless, manifestations of the continuing tradition of astronomical allusions in literature. Even judges have elevated their prose toward the Sun: "her whole being radiant with the sun-shine of the ineffable joys of youth." (Beers v. California State Life Ins. Co., 87 Cal. App. 440, 32.).

Writers and speakers frequently employ astronomical metaphors in everyday language as well as in poetry and song. The Sun, the Moon, and the stars are universal references readily compared to other objects, and references to these celestial objects increase the power and nuances of our language. Indeed, astronomical metaphors are so common and ingrained in our language that we often use them unconsciously, unaware of the underlying poetic device.

If a sunset to humankind's fascination with astronomy must eventually occur, it is still far in the future.

Glossary

a posteriori: Latin; a type of reasoning, from observation to theory, from facts or particulars to general principles, from effects to causes. Contrast with a priori. Synonymous with inductive and empirical. See *regular solids.*
a priori: Latin; a type of reasoning, from theory or principles without prior observations. Contrast with a posteriori. See *regular solids.*
acentric: not having a center.
AD: originally an abbreviation for Anno Diocletiana, the year of Diocletion, a year designation in a system of numbering years beginning with the beginning of Diocletian's reign in what is now AD (Anno Domini) 284. AD as Anno Domini counts years from what was supposed to be the birth of Christ. See *CE, BC, BCE.*
alignment: in a line with something.
anomalistic month: the time it takes the Moon to return to a point (such as the perigee or apogee) of its orbit. An anomalistic month is 27.55 days.
anomaly: something that is strange or unexpected; a discovery for which an investigator's paradigm has not prepared him or her.
anthropocentric universe: a worldview centered on human interests. The Copernican revolution saw a historical progression from belief in a small universe with humankind at its center to a larger and eventually infinite universe with the Earth not in the center. The physical geometry of our universe was transformed from geocentric and homocentric to heliocentric, and eventually to acentric. The psychological change was no less. Humans no longer commanded unique status as residents of the center of the universe, enjoying a privileged place. Nor was it likely that *Homo sapiens* were the only rational beings in the universe. Some might even question whether a good God sent an Adam and an Eve and a Jesus Christ only to Earth, or to every planet.
antiperistasis: Aristotle's convoluted explanation for how an arrow shot from a bow or a stone thrown from a hand continued in motion with continuing contact between the moved and the mover. Somehow air pushed forward by the arrow or the stone moved around to the rear of the arrow or stone and then pushed it from behind. See *impressed force.*
aphelion: the point in an orbit farthest from the Sun. *apo* = away from; *helio* = the Sun. See *perihelion.*
apogee: the point in an orbit farthest from the Earth. *gee* = the Earth. See *perigee.*
apsides: plural of apsis. See *apsis.*

apsis: the point in the orbit of a body where the body is neither approaching nor receding from another body about which it revolves. Elliptical orbits have two apsides, at the perigee and at the apogee.

archaeoastronomer: a person who practices archaeoastronomy. Aracheoastronomers find correlations between remains of ancient civilizations and positions of celestial objects.

archaeoastronomy: the study of ancient astronomy through investigations of archaeological remains.

archaeology: the scientific study of prehistoric and historic peoples and their cultures by analysis of their artifacts, inscriptions, monuments, and other physical remains.

Aristotelian physics: Aristotle held that the universe was divided into two parts, the terrestrial or earthly region, and the heavenly or celestial region. In the earthly region, all bodies were composed of combinations of the four basic elements: earth, fire, air, and water. Beyond the Moon, the heavenly bodies such as the Sun, the stars, and the planets were made of a fifth substance, the quintessence. A fundamental assumption of Aristotelian physics was that sublunary matter naturally moved toward its natural place in the center of the universe.

asterism: a star pattern that is not a constellation. Ursa Major is a constellation, but the Big Dipper, which is part of Ursa Major, is an asterism.

astrology: the practice of foretelling the future from the positions of celestial bodies, which are believed by some people to influence physical events and human affairs on the Earth. The need for astrologers to determine the positions of celestial objects was once an important source of patronage for astronomers, and a factor in the development of astronomy.

astrotheology: the eighteenth-century belief in the orderliness of the universe combined with a belief that determination of that order was an important theological, philosophical, and scientific endeavor.

atomism: a system of thought holding that all bodies are composed of minute, indivisible particles of matter, called atoms. Atomism as a philosophy originated with the ancient Greek philosopher Leucippus (first half of the fifth century BC) and his disciple Democritus (ca. 460–370 BC). They taught that all bodies are composed of atoms and the spaces between them, and that atoms are eternal, indivisible, infinite, and homogenous.

autumnal equinox: See *equinox.*

axis of rotation: the center around which something rotates. For the Earth, the axis of rotation is a line extending through the Earth from the north pole to the south pole. The Earth's axis of rotation is tilted at an angle of 23.5 degrees to the plane of the Earth's orbit around the Sun.

basilica: a church building with a central nave and two or four aisles formed by rows of columns; also a church afforded special privileges by the pope.

BC: a year designation in a system of counting calendar dates backwards from a date selected to represent the birth of Christ. See *AD, BCE, CE.*

BCE: Before Common Era, an alternative to AD; less specifically Christian, but still employing the birth of Christ as its end point. See *AD, BC, CE.*

big bang: a theory that the universe began about 15 billion years ago with a tremendous explosion, and that is has been continually expanding ever since.

bluestone: a bluish-gray sandstone used in the construction of Stonehenge. The Stonehenge bluestones came from Wales.

caduceus: the staff carried by Hermes, or Mercury, the messenger of the gods. Two snakes are entwined around the staff. A representation of this staff is sometimes mistakenly used as an emblem of the medical profession; Asclepius, the father of medicine, carried a staff with only one snake.
cairn: a pile of rocks. It may mark an alignment with a celestial body.
CE: Common Era, an alternative to AD; less specifically Christian, but still employing the birth of Christ as its starting point. See *AD, BC, BCE.*
celestial: referring to the sky or the heavens. The term is geocentric because it lumps together everything that is not the Earth.
celestial spheres: a fundamental part of the earth-centered astronomies of Plato, Aristotle, and Ptolemy. In their geocentric models, the stars and planets were carried around the Earth on spheres or circles.
chain: the Scottish economist Adam Smith (1723–1790), who initially believed that philosophical systems were mere inventions of the imagination to connect together otherwise disjointed and discordant phenomena of nature, was persuaded by the example of Isaac Newton's astronomy that there are real connecting principles or chains that bind nature. See *great chain of being.*
circle: a closed curve everywhere the same distance from its center. Ancient Greek philosophers deemed it the perfect geometrical figure. Supposedly, Plato (424–348 BC) set for astronomers the task of explaining the apparently irregular motions of the planets, Sun, and Moon as combinations of circular motions with constant speeds. See *save the appearances.*
circumference: the distance around a circle, or the outside line of the circle itself.
comet: a celestial body with a dense head and a vaporous tail. It is visible only if, and when, its orbit takes it near the Sun. Aristotle made a distinction between the corrupt and changing sublunary world (from the Earth up to the Moon) and the perfect, immutable heavens beyond. Tycho Brahe's observations of the comet of 1577 showed that it was above the Moon and moving through regions of the solar system previously believed filled with crystalline spheres carrying around the planets. Aristotle's world was shattered. See *crystalline sphere; new star.*
commentaries: ancient and medieval discussions or presentations based on books by Aristotle. Commentaries were neither modern nor scientific in spirit. Everyone knew what the questions were and what the answers would be. The goal was skillful presentation of known information, not discovery of new information. The medieval university curriculum was based partly on studying commentaries on Aristotle's works on logic and natural philosophy.
condemnation of 1277: necessary principles can result in truths necessary to philosophy but contradictory to dogmas of the Christian faith. In 1270, the Bishop of Paris condemned several propositions derived from the teachings of Aristotle, including the eternity of the world and the necessary control of terrestrial events by celestial bodies. In 1277, the pope directed the bishop to investigate intellectual controversies at the university. Within three weeks, over two hundred propositions were condemned. Excommunication was the penalty for holding even one of the damned errors. Some historians assert that the Scientific Revolution of the sixteenth and seventeenth centuries owed much to the condemnation of 1277. Though intended to contain and control scientific inquiry, the condemnation may have helped free cosmology from Aristotelian prejudices and modes of argument. But if so, why did scholars wait hundreds of years before repudiating Aristotelian cosmology?

conic section: a curve produced from the intersection of a plane with a right circular cone: circle, ellipse, hyperbola, and parabola. A book on conic sections by the ancient Greek mathematician Apollonius (ca. 262–190 BC) was recovered along with other ancient Greek classics in the early stages of the Renaissance. Perhaps when the mathematician and astronomer Johannes Kepler (1571–1630) casually jettisoned circular orbits and two thousand years of tradition, he was enabled by his understanding that the circle and the ellipse are both conic sections, neither geometrical figure more natural nor more perfect than the other.

constellation: a group of stars apparently related to each other in a pattern. Well-known constellations include Orion, Ursa Major (the Big Bear), and Ursa Minor (the Little Bear). These groupings are arbitrary, and the grouped stars generally have no physical relation to one another. Indeed, they are often at very different distances from the Earth. Constellations are convenient for describing sky locations. There are eighty-eight standard constellations.

cosmology: a subset of astronomy dealing with the origin, structure, and evolution of the universe.

Council of Trent: a council of the Roman Catholic Church held at Trent in northern Italy between 1545 and 1563 with an object of definitively determining Church doctrines in response to Protestantism.

Counter-Reformation: Roman Catholic efforts beginning in the sixteenth century to oppose the Protestant Reformation and reform the Catholic Church.

crisis: in science, a period of great uncertainty when an anomaly is judged worthy of concerted scrutiny yet continues to resist increased attention, and there are repeated failures to make the anomaly conform. A crisis often ends in large-scale paradigm destruction and major shifts in the problems and techniques of normal science. As well as scientific observations, external social, economic, and intellectual conditions can transform an anomaly into a crisis.

crystalline sphere: in ancient Greek and medieval European cosmology, solid crystalline spheres provided a physical structure for the universe, and carried the Sun, Moon, planets, and stars around the Earth. When Copernicus moved the Earth out of the center of the universe, the spheres were rendered redundant, and Tycho Brahe's observations of the comet of 1577 shattered the spheres. See *comet.*

cuneiform writing: an ancient Babylonian writing system using wedge-shaped signs made by pressing a sharpened stylus or stick into a soft clay tablet, often about the size of a hand. The tablets were dried or baked hard, preserving the contents, in some instances astronomical observations.

deduction: a logical derivation from theory of expected phenomena. An inference of the sort that if the premises are true, the conclusion necessarily follows. See *hypothetico-deductive method*; *induction.*

deferent: in ancient Greek geometrical astronomy, a large circle rotating at a constant speed around its center (coinciding with the Earth) and carrying around on its circumference the center of a smaller rotating circle (an epicycle), which, in turn, carried a planet on its circumference. See *eccentric*; *epicycle.*

divination: the practice of determining the hidden significance of events or foretelling the future by interpreting omens, which ranged from chicken entrails to astronomical data and events.

draconic month: the time for the Moon to travel around its orbit and return to the same node. A draconic month is 27.2 days. See *lunar node.*

draconic year: the time for the Sun to travel around its apparent orbit and back to the same lunar node. See *lunar node.*
earthshine: sunlight reflected from the Earth. Near new moon, the portion of the Moon shadowed from direct sunlight is slightly brightened by sunlight reflected from the Earth.
eccentric: a deferent circle with its center offset from the Earth. See *deferent.*
eccentricity: a measure of how far an ellipse deviates from a circle. A circle is a special case of an ellipse, with zero eccentricity. As an ellipse becomes more flattened, its eccentricity approaches one. Thus all ellipses have eccentricities between zero and one.
eclipse: a partial or complete temporary blockage of light by an intervening body. A solar eclipse occurs when the Moon comes between the Earth and the Sun. A lunar eclipse occurs when the Earth is between the Moon and the Sun.
ecliptic: the apparent path of the Sun among the stars; the intersection of the plane of the Earth's orbit around the Sun with the celestial sphere. See *zodiac.*
ellipse: a plane curved figure in which the sums of the distance of each point on the periphery from two fixed points, the foci, are equal; the closed curve generated by a point moving in such a way that the sum of the point's distances from two fixed points is a constant. Kepler's first law states that the planets' orbits are ellipses. See *Kepler's first law.*
empirical: originating in or based on observation or experience.
ephemeris: a tabular statement of the location of a celestial body at regular intervals. Ancient Babylonian astronomers seemingly were content with tables of predicted celestial positions, and they did not construct geometrical models to reproduce the motion of celestial bodies. Nor did they express concern about the causes of the motions nor curiosity about the physical composition of the celestial bodies, at least not in clay tablets found and studied so far.
epicycle: in ancient Greek geometrical astronomy, a small circle rotating at a constant speed around its center and carrying around a planet on its circumference. The center of the epicycle was carried around on the circumference of a larger rotating circle, the deferent.
equant: a geometrical invention of the ancient astronomer Ptolemy necessary to save the phenomena. Uniform angular motion, previously defined as cutting off equal angles in equal times at the center of the circle, was now taken with respect to this new point not at the center of the circle. The equant point was a questionable modification of uniform circular motion, and Copernicus would condemn the equant point as an unacceptable violation of uniform circular motion. See *save the appearances.*
equator: the great circle around the Earth's surface, defined by a plane passing through the Earth's center and perpendicular to its axis of rotation.
equinox: either of the two points (the vernal, or spring, equinox, about March 21; and the autumnal, or fall, equinox, about September 22) on the celestial sphere where the plane of the Earth's orbit around the Sun and the plane of the Earth's equator intersect. At these times of the year, the length of day and night are equal (twelve hours) everywhere on the Earth.
ex post facto argument: made after the fact or observation upon which it is based. Also a theory modified to bring it into agreement with new facts. Such theories carry less psychological conviction than do those predicting previously unknown phenomena, even if the strict logic of the two situations is equally compelling.

extrapolate: to project or extend known data into an area not known, so as to arrive at a conjectural knowledge of the unknown area.
februa: goat-skin thongs used in an ancient Roman fertility feast by two young men racing around the city walls and snapping the thongs (purifiers) at everyone they met. Source of the name of the month of February.
feng shui: a Chinese practice in which the placement of objects and the arrangement of space are intended to achieve harmony within the environment.
fixed stars: to early observers, stars that appeared motionless relative to each other, in contrast to moveable stars such as the Sun, Moon, and planets.
foundational: leading to. Kepler's observations did not lead to Newton's derivation of the inverse square law of gravity. Rather, establishment of the concept of universal gravitation enshrined Kepler's three laws among the great achievements of science. Although not foundational, Kepler's laws were important in the acceptance of Newton's theory.
full moon: the phase of the Moon when its entire disc is seen illuminated with light from the Sun. This occurs when the Moon is opposite the Earth from the Sun. See *new moon.*
geocentric: a view of the solar system, held by Aristotle and Ptolemy with the Earth in the center. See *heliocentric.*
geoglyph: a large drawing on the ground produced either by arranging stones or earth to create a positive geoglyph, or by removing rocks or soil, and thus exposing fresh ground, to create a negative geoglyph.
gravity: Isaac Newton explained planetary motions by the inverse square power of a mysterious entity called gravity, but he was unable to explain the cause of this power. He argued for setting aside the question of what gravity is, and to be content with a mathematical description of its effects: "And to us it is enough that gravity does really exist, and act according to the laws which we have explained, and abundantly serves to account for all the motions of the celestial bodies, and of our sea." (Newton, *Principia*)
great chain of being: in philosophical thought, it linked God to man to lifeless matter in a world in which every being was related to every other in a continuously graded, hierarchical order. Governmental order reflected the order of the cosmos, and thus social mobility and political change would be crimes against nature.
Gregorian calendar: a reform of the Julian calendar decreed by Pope Gregory XIII in 1582. It is now the most widely used calendar in the world.
Gregorian chant: a form of monophonic liturgical chant that accompanied rituals in the Western Church.
handbook tradition: Greek science was distilled into handbooks, and through this medium became known to Latin readers. Ptolemy wrote his *Almagest* after the incorporation of Greek science into handbooks had ceased, and the *Almagest* consequently was lost to Latin readers in Western Europe for many centuries.
harmony: an aesthetically pleasing combination of parts. Believing that God established nothing without geometrical beauty, Kepler compared the intervals between planets with harmonic ratios in music. See *symmetry.*
heliacal: occurring near the Sun, especially the risings and settings of a star or planet close to the rising and setting of the Sun. The heliacal rising of a star occurs when the star first becomes visible above the eastern horizon at dawn, after being hidden below the horizon or by the brightness of the Sun. The heliacal setting

occurs when the star is no longer visible in the sky at dawn because it has already set below the western horizon.
heliacal rising: See *heliacal.*
heliacal setting: See *heliacal.*
heliocentric: a view of the solar system with the Sun in the center; the Copernican universe. See *geocentric.*
Humanism: a cultural and intellectual movement from roughly 1300 to 1600, centered on the recovery of ancient Greek knowledge and an approach to ancient learning for its own sake rather than for its utility to Christianity. Humanism's basic philosophical outlook included a belief that the ultimate authority is human reason rather than some outside source. With Humanism came a renewed interest in Plato. Neo-Platonism, which also incorporated Neo-Pythagorean thought, included belief in the possibility and importance of discovering simple arithmetic and geometric regularities in nature, and also veiwed the Sun as the source of all vital principles and forces in the universe. Copernicus would rely on the recovery of Ptolemy's mathematical astronomy for both its geometrical techniques and its philosophical human values. Eventually, what had begun as a rebirth or recovery of old knowledge mutated into the creation of new knowledge.
hypothetico-deductive method: first, an hypothesis is postulated; then a prediction is deduced from the hypothesis; finally, observations are made to determine if the phenomenon deduced from the hypothesis exists. Note that hypotheses can be refuted, but not proved.
Ides: in the Roman calendar, the days in the middle of the month. They were originally the days of the full moon, but later, when longer months were used, became the thirteenth or fifteenth day. See *Kalends*; *Nones.*
igneous: rocks formed by solidification of cooled magma. Compare with sedimentary rocks formed of hardened lake-bed sediment, and with metamorphic rocks, which are igneous or sedimentary rocks transformed by great heat or pressure.
impetus: a moving force. See *impressed force*; *inertia*; *momentum.*
impressed force: an alternative to Aristotelian explanations of motion, which maintained contact between the mover and the moved (see *antiperistasis*). Some sort of incorporeal motive force could be imparted by a projector to a projectile. Planetary movements could be attributed to an impetus impressed by God at the creation. See *impetus*; *momentum.*
induction: the reasoning process in which observations (phenomena) somehow are followed by the formation of theories (propositions). Science does not correspond to the inductive model in which all facts are collected and then inevitable theories are inevitably induced. This model fails because of an infinite number of possible observations; true inductive science would never advance beyond the infinite period of fact gathering to the stage of theory formulation. Modern science increasingly deviates from the inductive model. Theories increasingly are used to suggest observations of potential significance. Intuition and hypotheses are essential catalysts in linking together observations and in guiding searches for new data. In contrast to deduction, in induction it is possible for the premises to be true but the conclusion to be false. This is why it is possible logically to refute a theory, but not to prove a theory. See *deduction.*
inertia: the tendency of a body at rest to remain at rest; or if in motion, to remain in motion. See *impetus*; *impressed force*; *momentum.*

inferior planet: a planet, either Mercury or Venus, lying between the Sun and the Earth. The inferior planets can never be seen on the opposite side of the Earth from the Sun. See *opposition*; *superior planet*.
instrumentalism: in the instrumentalist view, scientific theories are mathematical fictions relating observed phenomena, with no question of theories being true or even probable explanations of nature. For instrumentalists, it is enough that a scientific theory yield predictions corresponding to observations. Theories are simply calculating devices. See *nominalism*; *realism*.
intelligent design: a religious doctrine holding that the complexity of the universe is such that it can only be the result of purposeful design.
intercalation: the insertion of an extra (leap) day or days, week, or month into a calendar year to bring that time period into congruity with the solar year. Calendar years have whole numbers of days, but the solar year does not, and they customarily are reconciled by varying the number of days in different calendar years. See *leap day*.
Julian calendar: Julius Caesar's 46 BC reform of the Roman calendar. It has a twelve-month year with a leap day added to February every four years.
Kalends: the first day of the month in the Roman calendar. See *Ides*; *Nones*.
Kepler's first law: the planets move in elliptical orbits with the Sun at one focus. Announced in Kepler's *Astronomia nova* [New Astronomy] of 1609, and discovered after his "second" law. Kepler later attributed to Divine Providence the fact that Tycho Brahe had set him to work on Mars's orbit. It alone among the planetary orbits then known deviated sufficiently from a circle to force Kepler to reject a circular orbit. Next Kepler found that the discrepancies between Tycho's observations and circular and oval orbits were equal and opposite, and an elliptical orbit fit in between the circular and the oval. See *ellipse*.
Kepler's second law: or law of equal areas. Also announced in Kepler's *Astronomia nova* [New Astronomy] of 1609, and discovered before his "first" law. The radius vector, the line from the Sun to the planet, sweeps out equal areas in equal times. If the areas of any two segments are equal, then the times for the planet to travel between the points on the orbit defining the two segments are also equal. Thus the distance of a planet from the Sun is inversely related to its orbital velocity: as the distance increases, the velocity decreases. That the planets move faster the nearer they are to the Sun had already been cited by Copernicus as a celestial harmony; Kepler now found further harmony in a quantitative formulation of the relationship. See *harmony*.
Kepler's third law: or harmonic law. Announced in Kepler's *Harmonices mundi* [Harmonies of the World] in 1619. Among many propositions in this book on cosmic harmonies detailing various planetary ratios was the statement that the ratio of the mean movements of two planets is the inverse ratio of the 3/2 power of the spheres. Kepler's few readers could scarcely have guessed that this particular harmony would later be singled out for acclaim, while all the other numerical relationships in the book would be discarded as nonsense, tossed into the garbage can of history. The law is also expressed as a ratio between two planets (A and B) going around the Sun (and also between two satellites going around a planet): the ratio of the periods squared is equal to the ratio of the distances cubed: (period A/period B) squared = (distance A/distance B) cubed. See *harmony*.
Koran: traditional English spelling for Qur'an, the central religious text of Islam.

kudurru: a stone used by the Babylonians with inscribed details of land grants. Kudurrus are often called boundary stones, suggesting that they marked the edges of property, but the surviving ones were found in temples, where it appears they were kept in the manner that a county records office now keeps property records.
latitude: a measurement of angular distance from the equator. The equator is numbered as latitude zero; the north pole as 90 degrees north and the south pole as 90 degrees south.
leap day: a day added to a calendar year. See *intercalation.*
logarithm: the exponent of the power to which a base number must be raised to equal a given number. The logarithm L to the base 10 of the number X is defined by the equation X equals 10 to the L power. Though it may no longer be impressive or seem like a big deal to owners of electronic calculators, the introduction of logarithms early in the seventeenth century reduced the otherwise lengthy process of multiplication to simple addition, and thus probably doubled Kepler's productivity and his working lifetime.
longitude: a measurement of angular distance from the prime meridian, an imaginary line going from pole to pole and through Greenwich, England. The equivalent line on the opposite side of the Earth is longitude 180 east and 180 west.
lunar calendar: a calendar in which the cycle of the Moon's phases is the basic measuring unit. The Islamic calendar is lunar.
lunar node: a point on the plane of the Moon's orbit where the plane of the Moon intersects the plane of the apparent path of the Sun around the Earth. See *Saros cycle.*
lunation: the cycle from new moon to new moon, about 29.5 days.
medicine wheel: a circle of stones with astronomical alignments. Wheels have been found in Wyoming, South Dakota, Montana, Alberta, and Saskatchewan. The first wheel was identified on a ridge of Medicine Mountain in Wyoming's Big Horn Range.
megalith: a very large, usually rough stone, especially one used by a prehistoric culture as a monument or building block.
mercury: a metallic element, free-flowing at room temperature, and named after the planet Mercury. Alchemists believed that mercury was the first matter from which other metals were formed, and they thought that different metals could be produced by varying the quantity of sulfur mixed with mercury. Thus mercury was required for the transmutation of base metals into gold. Mercury was also used as a treatment for syphilis, a use that may be responsible for the proverb: "One night with Venus; seven years with Mercury." Mercury is acutely hazardous as a vapor and in the form of its water-soluble salts, which corrode membranes of the body. Mercury poisoning can cause irreversible brain, liver, and kidney damage, and is potentially fatal.
Metonic cycle: an astronomical cycle, whose discovery is credited to the ancient Greek astronomer Meton (fifth century BC). Nineteen solar years (6,939.602 days) are nearly the equivalent of 235 lunar months (6,939.688 days). This fact is the basis of a system for reconciling solar and lunar calendars.
Milky Way: a band of light in the night sky; also the galaxy that contains our solar system.
Mithraism: a mystery religion based on the worship of Mithra.
momentum: a force of motion, the product of a body's mass times its linear velocity. See *impetus*; *impressed force*; *inertia.*

motet: a medieval, unaccompanied choral composition with different texts sung simultaneously over a Gregorian chant.
mythology: a set of stories, traditions, or beliefs held by a group of people.
nebular hypothesis: a theory about the origin of the solar system. It began as a large nebula, just dense enough to have begun contracting from the force of its own gravity. As the nebula collapsed, it heated up, its spin increased, and it flattened.
Neo-Platonism: a philosophical movement between 1300 and 1600, featuring a renewed interest in Plato. A Neo-Pythagorean element within Neo-Platonism emphasized the possibility and importance of discovering simple arithmetic and geometric regularities in nature, and also viewed the Sun as a source of vital principles and forces in the universe. See *Neo-Pythagoreanism*; *Humanism.*
Neo-Pythagoreanism: a medieval and Renaissance philosophy which viewed Pythagoras rather than Plato as the central figure in the ancient Greek philosophical tradition. As interest in Plato grew, Neo-Platonism replaced Neo-pythagoreanism. See *Neo-Platonism*; *Humanism.*
new moon: the phase of the Moon when none of its disc is seen illuminated with light from the Sun. This occurs when the Moon is between the Earth and the Sun. In Islamic and Hebrew cultures, the new moon is the first sighting of the crescent moon. See *full moon.*
new star: a nova or supernova, a star that explodes and increases hundreds of millions of times in brightness. The famous nova of 1572 struck a blow against the Aristotelian worldview, in which there could be no change in the heavens. See *comet.*
Nicene Creed: a unified statement of Christian belief adopted at the First Council of Nicaea in 325.
node: See *lunar node.*
nominalism: the doctrine that scientific theories are working hypotheses in agreement with observed phenomena, but not necessarily true descriptions of the world. The nominalist thesis, developed in the 1300s, conceded the divine omnipotence of Christian doctrine, but at the same time freed natural philosophy from religious authority. Nominalism stripped of its religious context became instrumentalism, and nominalists have been favorably characterized as forerunners of modern, philosophically sophisticated instrumentalists. See *instrumentalism.*
Nones: the day of the month in the Roman calendar in which the Moon reached its first quarter. When the Romans divorced the month from the phases of the Moon, Nones became either the fifth or the seventh day. See *Ides*; *Kalends.*
normal science: the continuation of a research tradition, seeking facts indicated by theory to be of interest.
nova: See *new star.*
opposition: a celestial object in opposition is located on the opposite side of the Earth from the Sun. The inferior planets, Mercury and Venus, can never be at opposition because of the geometry of the situation. Hence in the Ptolemaic Earth-centered model, the speeds around the Earth of the Sun and an inferior planet (Mercury or Venus) must be nearly matched to keep the planet and the Sun in approximately the same angular direction as seen from the Earth. In the Copernican model, an inferior planet is always at a small angle from the Sun. Nothing further is required of the model builder to obtain this result; it is a natural, inherent, and inevitable consequence of the model.
paradigm: a universally recognized achievement temporarily providing model problems and solutions to a community of practitioners. A paradigm informs

scientists about the entities that nature contains and about the ways in which these entities behave.

parallax: the difference in apparent direction of an object seen from different places; a measure of distance. See *stellar parallax.*

Pascal Moon: the new moon whose fourteenth day follows the vernal or spring equinox.

perfect solids: See *regular solids.*

perigee: the point in an orbit nearest the Earth. *peri* = near; *gee* = the Earth. See *apogee.*

perihelion: the point in an orbit nearest the Sun. *helio* = the Sun. See *aphelion.*

periodicity: the quality of occurring at regular intervals; also the interval itself.

petroglyph: an image carved or engraved into rock.

picaresque: a type of fiction dealing with the episodic adventures of an often roguish protagonist.

Platonic solids: See *Pythagorean solids*; *regular solids.*

precursitis: an imaginary disease involving the unconscious assumption that ancient scientists were working on the same problems and using the same methods as modern scientists are. Hence the search for ancient precursors of the observations and theories now acclaimed in textbooks, and the myopic result: a chronology of cumulative systematized positive knowledge.

principle of plenitude: the idea that no genuine potentiality of existence could fail to exist and that the extent and abundance of creation must be as great as possible.

prognostications: predictions based on signs or omens.

Pythagorean solids: See *Platonic solids*; *regular solids.*

Pythagoreanism: metaphysics and philosophy of the ancient Greek philosopher Pythagoras (ca. 580–ca. 500 BC). From their observation that attributes and ratios of musical scales are expressible in numbers, Pythagoreans concluded that the heavens were composed of musical scales and numbers, and the planets' movements produced harmonious music. Pythagoreans also attempted to understand qualities, including justice and goodness, in terms of numerical ratios. Their mathematics and astronomy is sometimes characterized as primarily a religious exercise. See *Neo-Pythagoreanism.*

quadrant: an instrument for measuring angles, particularly the angle subtended by two celestial objects as viewed from the Earth.

qualitative: of, relating to, or involving quality or kind, as opposed to number or quantity. Not precise. See *quantitative.*

quantitative: involving measurement and numbers. Throughout the eighteenth century, Newtonians added quantitative success to quantitative success, mathematically explaining phenomena as the result of an inverse square force, while rival Cartesians failed to reconcile their vortex theory in any numerical detail with Kepler's mathematical laws of planetary motion.

Qur'an: current preferred English spelling for Koran, the central religious text of Islam.

realism: an insistence that scientific theories are descriptions of reality. Dogmatic realists insist on the truth of a theory. Critical realists concede a theory's conjectural character without necessarily becoming instrumentalists. A disappointed realist may appear to be a local instrumentalist with regard to a particular failed theory retaining instrumental value, but is far from becoming a global instrumentalist. See *instrumentalism.*

regular solids: geometrical solids, each with all sides equal, all angles equal, and all faces identical. There are five regular, or perfect, or Platonic, or Pythagorean solids; no more, no less. They are the tetrahedron (four triangular sides), cube (six square sides), octahedron (eight triangular sides), dodecahedron (twelve pentagonal sides), and icosahedron (twenty triangular sides). Kepler believed that it was more than coincidence that there were six planets (then known) with five intervals between them, corresponding to the five regular solids, and he proclaimed this discovery in 1596 in his *Mysterium cosmographicum* [Cosmic Mystery]. See *Platonic solids; Pythagorean solids.*

Renaissance: the rebirth (re-nascence) of Greek classical culture. Beginning in Italy in the fourteenth century, Renaissance Humanists looking to the past for knowledge from a higher civilization facilitated a rebirth of Greek philosophy and values. Inconsistencies within individual ancient works and between different authors, and discrepancies in the sciences between classical theory and contemporary observation, initially were attributed to defects in transmission and translation. Eventually, however, what had begun as a rebirth or recovery of old knowledge mutated into the creation of new knowledge. Obviously any ending date for the Renaissance is arbitrary; some choose 1632 and the trial of Galileo.

revolution-making: as opposed to revolutionary. The historian and philosopher of science Thomas Kuhn (1922–1996) argued that Copernicus's work was almost entirely within an ancient astronomical tradition, and hence not revolutionary, but it contained a few novelties that would lead to a scientific revolution, and hence was revolution-making. See *revolutionary; scientific revolution.*

revolutionary: involving a sudden, complete, or marked change. See *revolution-making; scientific revolution.*

Saros cycle: a cycle involving the periodicity and recurrence of eclipses. The cycle is 6,585.3 days (18 years, 11 days, 8 hours) long. It was known to Babylonian astronomers as a period when lunar eclipses repeat themselves, but the cycle is applicable to solar eclipses as well. The cycle arises from a harmony among three of the Moon's orbital periods: a synodic month (new moon to new moon) is 29.53059 days (29 days, 12 hours, 44 minutes); a draconic month (node to node) is 27.21222 days (27 days, 5 hours, 6 minutes); and an anomalistic month (perigee to perigee) is 27.55455 days (27 days, 13 hours, 19 minutes). One Saros is equal to 223 synodic months, to 242 draconic months, and to 239 anomalistic months (plus or minus a couple of hours).

sarsen: any of numerous large sandstone blocks found in south-central England. Sarsens are also called Druid stones, because the Druids were at one time thought responsible for stone structures such as Stonehenge. The sarsens used for Stonehenge came from Marlborough Downs, eighteen miles to the north.

save the appearances: in the context of ancient Greek geometrical astronomy, to devise a system of uniform circular motions that reproduced the observed phenomena, or appearances. The guiding theme or paradigm of Greek planetary astronomy was attributed to Plato, and set for astronomers the task of explaining the apparently irregular motions of the planets, Sun, and Moon using combinations of circular motions with constant speeds of rotation.

Scholasticism: a fusion of Aristotelianism and Christian ideas permeating thought in Western Europe between roughly 1200 and 1500, especially in universities. Next came Renaissance Humanism and a renewed interest in Plato.

Scientific Revolution: an extraordinary episode, during which scientific beliefs, values, and worldviews are abandoned and ruling paradigms are replaced by incompatible or incommensurable new paradigms. The Ptolemaic and Copernican systems were incompatible; they predicted different results, such as the appearance of the phases of Venus, but were judged by the mutual standard of how well each saved the phenomena with systems of uniform circular motions. Descartes' vortex system and Newton's gravity were more incommensurable than incompatible; they had different goals and different determinations of success or failure. Descartes insisted on an explanation of the cause of gravity, while Newton abandoned that quest and argued instead that it was enough that gravity acted according to an inverse square force law and accounted for all the motions of the celestial bodies. Some historians accept the replacement of one theory with a second, incompatible theory as a revolution. Other historians withhold the word *revolution* for the replacement of one worldview with another worldview so incommensurable that rival scientists cannot agree on common procedures, goals, and measurements of success.

sexagesimal number system: based on the number sixty. Our time system of sixty minutes in an hour and sixty seconds in a minute is an example. Ancient astronomical positions were reported as 28, 55, 57, 58, with each succeeding unit representing so many 60ths of the preceding unit: 28 degrees, 55 minutes, 57 seconds, and so forth.

sidereal month: the period of revolution of the Moon with respect to the stars. See *synodic month.*

solar year: also tropical year. The length of time for the Sun as viewed from the Earth to return to the same position relative to the seasons on the Earth. A solar year is 365 days, 48 minutes, 2.9 seconds.

solstice: either of two times of the year when the Sun is at its greatest distance from the celestial equator. The summer solstice in the northern hemisphere occurs about June 21 and the winter solstice occurs about December 21. The summer solstice is the longest day of the year and the winter solstice is the shortest. In the United States, the summer solstice is considered the first day of summer; in Europe, the summer solstice is Midsummer Day.

Star Trek: a popular and influential science fiction television series that takes place in the twenty-third century and features a spaceship that travels through the universe, and the crew encounters new life forms and new civilizations.

steady state: a theory about the universe with new matter continuously created while the universe expands. See *big bang.*

stellar: having to do with stars; also, the idea of outstanding, as in "It was a stellar performance."

stellar parallax: the angle subtended by the radius of the Earth's orbit at its distance from a star; the angle that would be subtended by one astronomical unit (the mean distance of the Earth from the Sun) at the distance of the star from the Sun. Stellar parallax was predicted from Copernicus's heliocentric theory, but at that time was too small to detect because the stars are at great distances from the Earth. Not until the 1830s, with instruments vastly better than the best in Copernicus's time, was stellar parallax first measured. See *parallax.*

step function: some Babylonian astronomers employed mathematical tables in which the solar velocity was represented as constant for several months, after which the Sun proceeded with a different constant speed for several more months, before reverting to the initial velocity and remaining at that speed for several more months. Graphed, the motion would look like a series of steps up and down, and is called a "step" function, although the Babylonians are not known to have used graphs. See *zigzag function.*

sublunary: below the Moon. See *Aristotelian physics.*

summer solstice: See *solstice.*

sunspot: a dark spot appearing on the face of the Sun, associated with magnetic fields, and with a twenty-two-year cycle from minimum to maximum and back to minimum activity.

superior planet: a planet lying beyond the Earth outward from the Sun (Mars, Jupiter, etc.). See *inferior planet.*

supernova: the explosion when a massive star exhausts its fuel and collapses, resulting in a sharp increase in brightness followed by gradual fading. The peak light output of a supernova explosion can outshine a galaxy. See *new star.*

syllogism: a form of deductive reasoning established by Aristotle. Certain things being stated, something other than what is stated follows of necessity from their being so. An example of a syllogism is: All organisms are mortal; all men are organisms; therefore, all men are mortal. Demonstrative syllogisms derive facts already known, not new facts. See *deduction.*

symmetry: harmonious arrangement in size, form, and arrangement of parts, often involving a sense of beauty. Plato regarded heaven and the bodies it contained as framed by the heavenly architect with the utmost beauty; Ptolemy contemplated many beautiful mathematical theories; Copernicus found "an admirable symmetry in the Universe, and a clear bond of harmony in the motion and magnitude of the Spheres." See *harmony.*

synodic month: the period of revolution of the Moon with respect to the Sun. See *sidereal month.*

syzygy: when the Moon lies in a straight line with the Earth and the Sun, at opposition or conjunction. It looks like a good Scrabble word, but the normal Scrabble set has only two tablets with the letter *y* on them.

themata: underlying beliefs, values, and worldviews constraining or motivating scientists and guiding or polarizing scientific communities.

thought experiment: a device of the imagination used to investigate the nature of things.

tropical year: See *solar year.*

Tusi couple: an innovative combination of uniform circular motions producing motion in a straight line, devised by the Islamic astronomer Nasir al-din al-Tusi in the thirteenth century. See *save the appearances.*

uniform circular motion: a circular motion at a constant speed. An object moving in uniform circular motion covers the same linear distance in each unit of time.

vernal equinox: See *equinox.*

vortices: sometimes vortexes. Huge whirlpools of cosmic matter. According to the French philosopher and scientist René Descartes (1596–1650), our solar system was one of many vortices, its planets all moving in the same direction in the same plane around a luminous central body. Planets' moons were swept along by the planets' vortices.

winter solstice: See *solstice.*

woodhenge: a wooden structure similar to Stonehenge. There may have been a number of these, but wood does not survive as well as stone.

zigzag function: a sequence of numbers in a Babylonian table of astronomical positions, decreasing, increasing, and then decreasing again. So called after how it would appear on a graph. Note that the Babylonians are not known to have used graphs. See *step function.*

zodiac: an imaginary belt circling the heavens, extending about eight degrees on each side of the ecliptic (the intersection of the plane of the Earth's orbit around the Sun with the celestial sphere). Within the zodiac are the apparent paths of the Sun, Moon, and planets. The zodiac is divided into twelve sections of thirty degrees each. Each division contains a constellation, which at the time of the Babylonians corresponded with the twelve signs of the zodiac. However, because of the precession of the equinoxes, each division now contains the constellation west of the one from which it took its name. Zodiac has the same root as the words *zoo* and *zoological,* and appropriately so, since most of the signs of the zodiac are animals. See *ecliptic.*

Annotated Bibliography

BOOKS

Primary Sources

Aristotle. *Physica*, in *Aristotle's Physics, newly translated by Richard Hope, with an Analytical Index of Technical Terms* (Lincoln: University of Nebraska Press, 1961). Aristotle's physics dominated scientific thought in the Western world for nearly two thousand years.

———. *De caelo*. J. L. Stocks, translator, in W. D. Ross, ed., *The Works of Aristotle, vol. 2* (Oxford: Clarendon Press, 1930), reprinted, with mild emendations, in Jonathan Barnes, ed., *The Complete Works of Aristotle*, vol. 1 (Princeton, NJ: Princeton University Press, 1984). In *On the Heavens*, Aristotle develops ideas raised in his *Physics*.

Chaucer, Geoffrey. *The Canterbury Tales*. Available in numerous editions, and online.

Copernicus, Nicholas. *Commentariolus*, in Edward Rosen, ed., *Three Copernican Treatises: The Commentariolus of Copernicus, the Letter against Werner, and the* Narratio Prima *of Rheticus. Second ed., revised, with an annotated Copernicus bibliography 1939–1958* (New York: Dover Publications, 1959). See also Noel Swerdlow, "The Derivation and First Draft of Copernicus' Planetary Theory: A Translation of the *Commentariolus* with Commentary," *Proceedings of the American Philosophical Society, 117* (1973), 423–512. Copernicus's early astronomical thoughts, in English translation and with extensive notes.

———. *De revolutionibus orbium caelestium*. Copernicus's great work of 1543 on the revolutions of the celestial spheres. The preface and book are conveniently available in English translation in Michael J. Crowe, *Theories of the World from Antiquity to the Copernican Revolution* (New York: Dover Publications, Inc., 1990). Much of the preface and book one are also presented and commented extensively upon in Thomas Kuhn, *The Copernican Revolution: Planetary Astronomy in the Development of Western Thought*.

Fontenelle, Bernard le Bovier de. *Entretiens sur la pluralité des mondes*. In English translation with a useful introduction, notes, and bibliography in Bernard le Bovier de Fontenelle, *Conversations on the Plurality of Worlds, translation by H. A. Hargreaves, introduction by Nina Rattner Gelbert* (Berkeley: University of California Press, 1990). Also available at Littérature française en edition électronique, http.//www.scribd.com/doc/236572/Fontenelle-B-Entretiens-sur-la-pluralitie-des-mondes. Fontenelle imagined evening promenades in a garden with a lovely young marquise, discussing that the Moon was inhabited, that the planets were also inhabited, and that the fixed stars were other suns, each giving light to their own worlds.

Galilei, Galileo. *Sidereus nuncius.* In English translation in Stillman Drake, *Discoveries and Opinions of Galileo, translated with an introduction and notes by Stillman Drake* (Garden City, NY: Doubleday, 1957). Extensive historical introduction and English translations of Galileo's *Starry Messenger* on his telescopic discoveries of 1610, his 1613 letters on sunspots, and his *Letter to the Grand Duchess Christina* in 1615 discussing the relationship between science and religion. Drake is a major Galileo scholar, and all of his many books on Galileo are good reading. Equally excellent is Albert van Helden, *Sidereus Nuncius or The Sidereal Messenger Galileo Galilei, translated with introduction, conclusion, and notes by Albert van Helden* (Chicago: University of Chicago Press, 1989). Contains a good bibliography and discussion of Galileo's discovery of the phases of Venus.

———. *Dialogue Concerning the Two Chief World Systems—Ptolemaic and Copernican. Translated, with revised notes, by Stillman Drake, foreword by Albert Einstein, 2nd ed.* (Berkeley: University of California Press, 1967). See also Maurice A. Finocchiaro, *Galileo on the World Systems: A New Abridged Translation and Guide* (Berkeley: University of California Press, 1997).

Hesiod: Theogony; Works and Days; Shield. Translation, introduction, and notes by Apostolos N. Athanassakis, 2nd ed. (Baltimore: Johns Hopkins University Press, 2004). One of several available translations of these works providing sources for Greek mythology, farming, astronomy, and calendars.

Homer. *The Iliad. Originally translated by E. V. Rieu; revised and updated by Peter Jones with D. C. H. Rieu; edited with an introduction and notes by Peter Jones* (New York: Penguin Books, 2003). One of many translations of this classic account of the final year of the Trojan War.

Homer. *The Odyssey. Translated by E. V. Rieu; revised translation by D. C. H. Rieu, in consultation with Peter V. Jones* (New York: Penguin Books, 1991). One of many translations of this classic tale of Odysseus's (Ulysses') journey home after the Trojan War.

Kepler, Johannes. *Mysterium cosmographicum.* The *Cosmic Mystery* of 1596 is available in English translation in *Mysterium Cosmographicum* = The Secret of the Universe, A. M. Duncan, translator; introduction and commentary by E. J. Aiton; with a preface by I. Bernard Cohen (New York: Abaris Books, 1981).

———. *Astronomia nova.* The *New Astronomy* of 1609 is available in English translation in William H. Donahue, translator, *New Astronomy/Johannes Kepler* (Cambridge: Cambridge University Press, 1992).

———. *Harmonices mundi.* The *Harmonies of the World* of 1619 is available in English translation in *The Harmony of the World/by Johannes Kepler; translated into English with an introduction and notes by E. J. Aiton, A. M. Duncan, and J. V. Field* (Philadelphia: American Philosophical Society, 1997; *Memoirs of the American Philosophical Society, 209*).

———. *Epitome astronomiae Copernicanae.* Books IV and V of the *Epitome of Copernican Astronomy* of 1617–1621 are available in English translation in

Charles Glenn Wallis, translator, *The Harmonies of the World: V*, in *Great Books of the Western World, vol. 16: Ptolemy, Copernicus, Kepler* (Chicago: Encyclopedia Britannica, Inc., 1939); reprinted in *Epitome of Copernican Astronomy and Harmonies of the World* (Amherst, NY: Prometheus Books, 1991).

Newton, Isaac. *Philosophiae naturalis principia mathematica.* Among many English translations of the *Mathematical Principles of Natural Philosophy* of 1686 are: *Mathematical Principles of Natural Philosophy and His System of the World by Sir Isaac Newton*; translated into English by Andrew Motte in 1729. The translation revised and supplied with an historical and explanatory appendix, by Florian Cajori (Berkeley: University of California Press, 1946); *Newton's Principia: The Central Argument*; Translation, notes, and expanded proofs by Dana Densmore; Translations and diagrams by William H. Donahue (Santa Fe, NM: Green Lion Press, 1996); *The Principia: Mathematical Principles of Natural Philosophy. By Isaac Newton*; A new translation by I. Bernard Cohen and Anne Whitman assisted by Julia Budenz; Preceded by a Guide to Newton's Principia by I. Bernard Cohen (Berkeley: University of California Press, 1999).

Ovid. *The Metamorphoses of Ovid, translated by Mary M. Innes* (Baltimore: Penguin Books, 1955). One of many available translations of these classic tales, which are always a good read.

Rheticus. *Narratio prima.* The first published account, by a young professor of mathematics, Rheticus, of part of Copernicus's *De revolutionibus.* In English translation and with extensive notes, in Edward Rosen, ed., *Three Copernican Treatises: The Commentariolus of Copernicus, The Letter against Werner, and the Narratio Prima of Rheticus. Second ed., revised, with an annotated Copernicus bibliography 1939–1958* (New York: Dover Publications, 1959).

Shakespeare, William. *Plays, Sonnets, and Poems.* Available in numerous editions, and online.

Simplicius of Athens. *Commentary on Aristotle's* On the Heavens. Simplicius's description of the problem that Plato set for astronomers is available in English translation in Pierre Duhem, *To Save the Phenomena: An Essay on the Idea of Physical Theory from Plato to Galileo, translated by E. Doland and C. Maschler* (Chicago: University of Chicago Press, 1969). Some scholars doubt Simplicius's authority and question Plato's putative role in the development of planetary theory; see Bernard R. Goldstein, *Theory and Observation in Ancient and Medieval Astronomy* (London: Variorum Reprints, 1985). Simplicius's description of the failure of Eudoxus's theory to save the phenomena is available in Thomas Heath, *Aristarchus of Samos: The Ancient Copernicus. A History of Greek Astronomy to Aristarchus together with Aristarchus's Treatise on the Sizes and Distances of the Sun and Moon. A New Greek Text with Translation and Notes* (Oxford: Clarendon Press, 1913; New York: Dover Publications, 1981), pp. 221–223.

Smith, Adam. *An Inquiry into the Nature and Causes of the Wealth of Nations* (Edinburgh: 1776). Available in many editions. As Newton's *Principia* changed understanding of the physical world and Darwin's *Origin of Species* changed understanding of the biological world, so Smith's *Wealth of Nations* changed understanding of the economic world. For how Smith's work was inspired by Newton's, see Norriss S. Hetherington, "Isaac Newton and Adam Smith: Intellectual Links between Natural Science and Economics," in Paul Theerman and

Adele F. Seeff, *Action and Reaction: Proceedings of a Symposium to Commemorate the Tercentenary of Newton's* Principia (Newark: University of Delaware Press, 1993), pp. 277–291.

Theon of Smyrna. *Expositio rerum mathematicarum ad legendum Platonem utilium.* A handbook of citations from earlier sources on arithmetic, music, and astronomy intended to provide an exposition of mathematical subjects useful for the study of Plato. Excerpts in English translation with regard to the equivalence of the eccentric and epicyclic hypotheses are available in Pierre Duhem, *To Save the Phenomena: An Essay on the Idea of Physical Theory from Plato to Galileo, translated by E. Doland and C. Maschler* (Chicago: University of Chicago Press, 1969), pp. 8–11.

Secondary Sources

Aiton, E. J. *The Vortex Theory of Planetary Motions* (London: Macdonald & Co., 1972). On Descartes' vortex theory, the primary rival of Newton's theory of gravitation.

Allen, Richard H. *Star Names: Their Lore and Meaning* (New York: Dover Publications, 1963). Also available at http://penelope.uchicago.edu/Thayer/E/Gazetteer/Topics/astronomy/_Texts/secondary/ALLSTA/home.html. Discusses myths related to stars.

Aveni, Anthony. *Empires of Time: Calendars, Clocks, and Cultures* (New York: Basic Books, 1989). Somewhat philosophical account of the history of the Western calendar and comparisons with the Mesoamerican ones.

———. *Stairways to the Stars: Skywatching in Three Great Ancient Cultures* (New York: John Wiley & Sons, 1997). Scholarly discussion of Mesoamerican archaeoastronomy.

Bierhorst, John. *The Mythology of South America* (New York: Oxford University Press, 2002). Introduction to the indigenous mythology of seven regions of South America. See also his books on the mythology of North America and the mythology of Mexico and Central America.

Buchdahl, Gerd. *The Image of Newton and Locke in the Age of Reason* (London: Sheed and Ward, 1961). Small booklet on Newton's image and its effect on eighteenth-century thought and imagination.

Burland, Cottie. *North American Indian Mythology* (London: Paul Hamlin, 1965; New York: Tudor Publishing, 1965). Library of the world's myths and legends series. In many editions.

Cohen, I. Bernard. *The Birth of a New Physics, Revised and Updated* (New York: W. W. Norton, 1985). Expanded version of the 1960 original. An outstanding book showing clearly how Galileo, Kepler, and Newton changed Aristotle's physics.

Crowe, Michael J. *The Extraterrestrial Life Debate 1750–1900: The Idea of a Plurality of Worlds from Kant to Lowell* (Cambridge: Cambridge University Press, 1986). For earlier periods, see Dick, *Plurality of Worlds: The Extraterrestrial Life Debate from Democritus to Kant* (Cambridge: Cambridge University Press, 1982).

———. *Theories of the World from Antiquity to the Copernican Revolution* (New York: Dover Publications, Inc., 1990). Outstandingly clear presentation, developed from a college course.

———. *The Extraterrestrial Life Debate, Antiquity to 1915: A Source Book* (Notre Dame, IN: Notre Dame University Press, 2008).

Danielson, Dennis. *The First Copernicus: Georg Joachim Rheticus and the Rise of the Copernican Revolution* (New York: Walker & Co., 2006). Without Rheticus, it is unlikely that Copernicus would have published his *De revolutionibus.*

Dick, Steven J. *Plurality of Worlds: The Extraterrestrial Life Debate from Democritus to Kant* (Cambridge: Cambridge University Press, 1982). The issue of extraterrestrial life was debated as early as the ancient Greeks. From the eighteenth through the nineteenth century, see Crowe, *The Extraterrestrial Life Debate 1750–1900: The Idea of a Plurality of Worlds from Kant to Lowell.* For the twentieth century, see Dick, *The Biological Universe: The Twentieth-Century Extraterrestrial Life Debate and the Limits of Science.*

———. *The Biological Universe: The Twentieth-Century Extraterrestrial Life Debate and the Limits of Science* (Cambridge: Cambridge University Press, 1996). For earlier periods, see Dick, *Plurality of Worlds: The Extraterrestrial Life Debate from Democritus to Kant*; Crowe, *The Extraterrestrial Life Debate 1750–1900: The Idea of a Plurality of Worlds from Kant to Lowell.*

———, and James E. Strick, *The Living Universe: NASA and the Development of Astrobiology* (New Brunswick, NJ: Rutgers University Press, 2004).

Drake, Stillman. *Discoveries and Opinions of Galileo, translated with an introduction and notes by Stillman Drake* (Garden City, NY: Doubleday, 1957). Galileo's telescopic discoveries in his *Starry Messenger* of 1610, his 1613 letters on sunspots, and his *Letter to the Grand Duchess Christina* of 1615 discussing the relationship between science and religion; in English translation with historical introduction. Drake is a major Galileo scholar, and all of his many books on Galileo are good reading.

Dreyer, J. L. E. *History of the Planetary Systems from Thales to Kepler* (Cambridge: Cambridge University Press, 1906); 2nd ed., revised by William Stahl, with a supplementary bibliography, and retitled *A History of Astronomy from Thales to Kepler* (New York: Dover Publications, 1953). Still useful reading after a century.

Duncan, David Ewing. *Calendar: Humanity's Epic Struggle to Determine a True and Accurate Year* (New York: Avon Books, Inc., 1988). Readable history of the development of the Western calendar.

Feingold, Mordechai. *The Newtonian Moment: Isaac Newton and the Making of Modern Culture* (New York and Oxford: New York Public Library/Oxford University Press, 2004). The best single volume on Newton's effect on modern culture and thought. This is the companion volume to a library exhibition.

Ferngren, Gary B., ed. *The History of Science and Religion in the Western Tradition: An Encyclopedia* (New York: Garland Publishing, Inc., 2000). Interesting, if somewhat brief, essays.

Gingerich, Owen. *The Eye of Heaven: Ptolemy, Copernicus, Kepler* (New York: The American Institute of Physics, 1993). Informative essays on Ptolemy, Copernicus, and Kepler.

Goldsmith, Donald, and Tobias Owen. *The Search for Life in the Universe*, 3rd ed. (Sausalito, CA: University Science Books, 2001). Comprehensive introduction to bioastronomy; used in many college courses.

Grant, Edward. *Physical Science in the Middle Ages* (New York: John Wiley & Sons, 1971). Excellent introduction by a major scholar in this field.

———. *Science and Religion, 400 B.C. to A.D. 1550: From Aristotle to Copernicus* (Westport, CT: Greenwood Press, 2004). The history of relations between science and religion from the early centuries of Christianity through the influx of Greco-Arabic science and natural philosophy into Western Europe during the twelfth and thirteenth centuries. In the *Greenwood Guides to Science and Religion* series; see also Richard G. Olson, *Science and Religion, 1450–1900: From Copernicus to Darwin* (Westport, CT: Greenwood Press, 2004).

Graves, Robert. *The Greek Myths,* vols. 1 and 2 (Hammondsworth, Penguin Books, 1985–1986). Famous retelling of Greek myths. In many editions.

Guthke, Karl S. *The Last Frontier: Imagining Other Worlds, from the Copernican Revolution to Modern Science Fiction, translated by Helen Atkins* (Ithaca, NY: Cornell University Press, 1990). Intellectual history of the plurality of worlds in Western culture, from the ancient Greeks to the middle of the twentieth century.

Hawkins, Gerald S. in collaboration with John B. White. *Stonehenge Decoded* (Garden City, NY: Doubleday, 1965). Hawkins showed that stones at Stonehenge are aligned with various celestial phenomena.

Hetherington, Norriss S. *Encyclopedia of Cosmology: Historical, Philosophical, and Scientific Foundations of Modern Cosmology* (New York: Garland Publishing, Inc., 1993). Many major entries of over 5,000 words; also includes literature and religion.

———. *Planetary Motions: A Historical Perspective* (Westport, CT: Greenwood Press, 2006). Development of a science, and some of its cultural, historical, and intellectual interactions with civilizations that nurtured it.

Hoskin, Michael. *The Cambridge Illustrated History of Astronomy* (Cambridge: Cambridge University Press, 1997). Outstanding text and illustrations from an outstanding historian of astronomy.

James, Jamie. *The Music of the Spheres: Music, Science, and the Natural Order of the Universe* (New York: Springer-Verlag, 1993). Beginning with Pythagoras, how developments in music and science have been interrelated; succinct, even cursory.

Johnson, Francis R. *Astronomical Thought in Renaissance England: A Study of the English Scientific Writings from 1500 to 1645* (Baltimore: Johns Hopkins University Press, 1937; New York: Octagon Books, 1968). A pioneering study and now a classic.

Kerényi, C. *The Gods of the Greeks,* translated by Norman Cameron (New York: Grove Press, Inc., 1960). Analysis of Greek myths using some of the earliest available sources.

Koestler, Arthur. *The Sleepwalkers: A History of Man's Changing Vision of the Universe* (New York: Macmillan, 1959). The chapters on Kepler are also published separately, as *The Watershed* (Garden City, NY: Doubleday, 1960). A great read from one of the twentieth century's great writers.

Koyré, Alexandre. *From the Closed World to the Infinite Universe* (Baltimore: Johns Hopkins University Press, 1957). Changes in man's conception of his universe and his place in it.

———. *Newtonian Studies* (London: Chapman & Hall, 1965). Essays analyzing how scientific ideas are related to philosophical thought and also determined by empirical controls.

Kragh, Helge. *Cosmology and Controversy: The Historical Development of Two Theories of the Universe* (Princeton, NJ: Princeton University Press, 1996). Outstanding study of the big bang and steady state cosmological theories around the middle of the twentieth century; remarkable in both breadth and depth.

Kuhn, Thomas. *The Copernican Revolution: Planetary Astronomy in the Development of Western Thought* (Cambridge, MA: Harvard University Press, 1957). An outstanding analysis of relations between theory, observation, and belief in the development of Western astronomy to Copernicus, particularly within the framework of Aristotelian physics.

———. *The Structure of Scientific Revolutions* (Chicago: University of Chicago Press, 1962); 2nd ed., enlarged (Chicago: University of Chicago Press, 1970). The most influential book ever written on how science works.

Lindberg, David. *The Beginnings of Western Science: The European Scientific Tradition in Philosophical, Religious, and Institutional Context, 600 B.C. to A.D. 1450* (Chicago: University of Chicago Press, 1992). Outstanding introduction to the topic, and highly recommended.

Lloyd, G. E. R. *Early Greek Science: Thales to Aristotle* (New York: W. W. Norton, 1970) and *Greek Science after Aristotle* (New York: W. W. Norton, 1973). These two slim volumes provide a good introduction to Greek science.

Lockyer, Norman. *The Dawn of Astronomy: A Study of the Temple-worship and Mythology of the Ancient Egyptians* (London: Cassell, 1894). One of the first studies of alignments of temples with celestial phenomena.

———. *Stonehenge and Other British Monuments Astronomically Considered* (London: Macmillan, 1906; expanded 1909). One of the first studies of Stonehenge in an astronomical context.

Lovejoy, Arthur. *The Great Chain of Being: A Study of the History of an Idea* (Cambridge, MA: Harvard University Press, 1936). Traces the history of the idea of plenitude, that a good God created the universe full of all possible things, and continuity in nature.

Neugebauer, Otto. *Exact Sciences in Antiquity*, 2nd ed. (Providence, RI: Brown University Press, 1957; New York: Dover Publications, 1969). The major book on Babylonian astronomy.

———. *Astronomy and History: Selected Essays* (New York: Springer-Verlag, 1983).

Newton, Robert R., *The Crime of Claudius Ptolemy* (Baltimore: Johns Hopkins University Press, 1977). Argues that Ptolemy was the most successful fraud in the history of science. In Ptolemy's defense, see Owen Gingerich: "Was Ptolemy a Fraud?" in Gingerich, *The Eye of Heaven: Ptolemy, Copernicus, Kepler* (New York: American Institute of Physics, 1993).

Nicolson, Marjorie. *Newton Demands the Muse: Newton's* Optics *and the Eighteenth Century Poets* (Princeton, NJ: Princeton University Press, 1946). One of several studies by this author of the impact of science on the literary imagination.

———. *Voyages to the Moon* (New York: Macmillan, 1948). History of fictional space travels, mainly, but not all, to the Moon, up to the nineteenth century.

———. *Science and Imagination* (Ithaca, NY: Gold Seal Books, 1962).

North, John. *Chaucer's Universe* (Oxford: Clarendon Press; New York: Oxford University Press, 1988). Finds much astronomical allegory in Chaucer's works.

———. *Stonehenge: A New Interpretation of Prehistoric Man and the Cosmos* (New York: Free Press, 1996). The author's intent is "to discover certain patterns of intellectual and religious behavior through a study of archaeological remains that seem to have been deliberately directed in some way towards phenomena in the heavens."

———. *Cosmos: An Illustrated History of Astronomy and Cosmology* (Chicago: University of Chicago Press, 2008). Earlier versions: *The Fontana History of Astronomy and Cosmology* (London: Fontana Press, 1994); *The Norton History of Astronomy and Cosmology* (New York: W. W. Norton, 1995). Perhaps the best single volume history of astronomy, and authoritative. North's own scholarly research covers the time spectrum from Stonehenge through the Middle Ages to twentieth-century relativistic cosmology.

Ragep, F. Jamil. *Nasir al-Din al-Tusi's Memoir on Astronomy (al-Tadhkira fi 'ilm al-hay'a)*; two volumes (New York: Springer-Verlag, 1993). The major work on al-Tusi. See also, Ragep's "The Two Versions of the Tusi Couple," in David King and George Saliba, *From Deferent to Equant: A Volume of Studies in the Ancient and Medieval Near East in Honor of E. S. Kennedy* (New York: New York Academy of Sciences, 1987), pp. 329–356.

Rochberg, Francesca. *The Heavenly Writing: Divination, Horoscopy, and Astronomy in Mesopotamian Culture* (Cambridge: Cambridge University Press, 2004). On changing interpretations of the significance of Babylonian astronomy in the history and philosophy of science, especially Chapter 1, "The Historiography of Mesopotamian Science." Anything by Rochberg is highly recommended.

Ruggles, Clive. *Ancient Astronomy: An Encyclopedia of Cosmologies and Myth* (Santa Barbara, CA: ABC-CLIO, 2005). Over 300 entries, on famous ancient sites, myths, and themes, including cosmologies and calendars.

Saliba, George. *A History of Arabic Astronomy: Planetary Theories during the Golden Age of Islam* (New York: New York University Press, 1994). See also "Whose Science is Arabic Science in Renaissance Europe?" and illustrations for al-Tusi's astronomy on Saliba's Web site: http://www.columbia.edu/~gas1/project/visions/case1/sci.3.html.

Shklovskii, I. S., and Carl Sagan. *Intelligent Life in the Universe* (New York: Dell Publishing Co., 1966). Cosmology, origins of life, and search for extraterrestrial life. Carl Sagan expanded this Russian study; now somewhat outdated.

Small, Robert. *An Account of the Astronomical Discoveries of Kepler. A reprinting of the 1804 text with a foreword by William D. Stahlman* (Madison: University of Wisconsin Press, 1963). Chapter by chapter analysis with notes.

Sofaer, Anna. *Chaco Astronomy: An Ancient American Cosmology* (Santa Fe, NM: Ocean Tree Books, 2008). On astronomical alignments and the culture of the people who lived in the American Southwest between 850 and 1300.

Stephenson, Bruce. *Kepler's Physical Astronomy* (New York: Springer-Verlag, 1987). Presents a detailed examination of Kepler's *Astronomia nova* and argues persuasively that Kepler was guided by physical concerns to transcend traditional astronomy.

Sullivan, Walter. *We Are Not Alone: The Continuing Search for Extraterrestrial Intelligence*, rev. ed. (New York: McGraw-Hill, 1993). Best-selling account of searches for extraterrestrial life; now outdated.

Thoren, Victor E. *The Lord of Uraniborg. A Biography of Tycho Brahe* (Cambridge: Cambridge University Press, 1990). The best book on both Brahe's life and his science.

Voelkel, James R. *Johannes Kepler and the New Astronomy* (New York: Oxford University Press, 1999). In the *Oxford Portraits in Science* series, by top scholars and writers, on the personalities of scientists and the thought processes leading to their discoveries.

Westfall, Richard S. *Never at Rest: A Biography of Isaac Newton* (Cambridge: Cambridge University Press, 1980). The most complete book on Newton and his science. An abbreviated and more accessible version is Westfall, *The Life of Isaac Newton* (Cambridge: Cambridge University Press, 1993).

———. *The Construction of Modern Science: Mechanisms and Mechanics* (New York: John Wiley & Sons, 1971). On the new synthesis of science in the seventeenth century, from Kepler and Galileo to Newton.

WEB SITES

Ancient Observatories: Timeless Knowledge. NASA Web site with brief descriptions of observatories, ancient and modern; some links to other Web sites: http://sunearthday.nasa.gov/2005/locations/chaco.htm.

As the World Turned: A Reader on the Progress of the Heliocentric Argument from Copernicus to Galileo. Contains excerpts from Nicholas Copernicus, *De Revolutionibus;* John Dee, *The Mathematicall Praeface;* Robert Recorde, *The Castle of Knowledge;* Marcellus Palingenius Stellatus, *The Zodiake of Life;* Giordano Bruno, *The Ash Wednesday Supper; and* Galileo Galilei, *Dialogue Concerning the Two Chief World Systems:* http://math.dartmouth.edu/~matc/Readers/renaissance.astro/0.intro.html.

Calendar Reform. Essays on the history of calendar reforms and on current proposals for reform; links to other Web sites: http://personal.ccu.edu/mccartyr/calendar-reform.html#BB.

Calendars through the Ages. Explains the astronomical basis for calendars; briefly describes various calendars; recommends books on calendars: http://webexhibits.org/calendars/index.html.

Chaucer, Geoffrey. *The Canterbury Tales:* http://www.canterburytales.org/canterbury_tales.html.

Cosmic Journey: A History of Scientific Cosmology. The history of cosmology from ancient Greek astronomy to space telescopes, with more than 35,000 words and 380 illustrations: http://www.aip.org/history/cosmology.

eChaucer: Chaucer in the Twenty-First Century. Texts, translations, concordance, chronology, images, glossary, and links to other Web sites: http://www.umm.maine.edu/faculty/necastro/chaucer/chronology.asp.

Encyclopedia Mythica. Internet encyclopedia of mythology, folklore, and religion; includes an image gallery. The mythology section covers six geographical regions: Africa, the Americas, Asia, Europe, the Middle East, and Oceania: http://www.pantheon.org.

Exhibits Online. The History of Science Department of the University of Oklahoma makes available from the university's extensive history of science collections digitized images from particular books and of particular individuals: http://hsci.cas.ou.edu/exhibits/exhibit.php?exbid=1.

Famous Trials; Trial of Galileo Galilei, 1633. Includes chronology, scriptural references, the text of Galileo's *Dialogue Concerning the Two Chief World Systems – Ptolemaic and Copernican* in English translation, 1616 admonition, 1633 depositions, Galileo's defense, papal condemnation, Galileo's recantation, selected letters, images and maps, biographical sketches, bibliography, and links: http://www.law.umkc.edu/faculty/projects/ftrials/galileo/galileo.html.

The Galileo Project. Hypertextual information on the life and work of Galileo and on the science of his time; biographical sketches of hundreds of members of the scientific community during the sixteenth and seventeenth centuries; information about the Inquisition and biographical information of important church figures; history of the calendar; glossary; bibliography: http://galileo.rice.edu/Catalog/NewFiles/digges_leo.html.

Historical Celestial Atlases on the World Wide Web. Listing of virtual exhibitions and digital editions of celestial atlases and maps; brief list of books on the history of constellations and star names. http://www.phys.uu.nl/%7Evgent/celestia/celestia.htm.

International Center for Archaeoastronomy. Online home of the *Archaeoastronomy and Ethnoastronomy News*; with links to other Web sources on archaeology and astronomy: http://www.wam.umd.edu/~tlaloc/archastro/index.html.

Isaac Newton and Thomas Jefferson. A talk at Monticello on the effects Newton had on the Enlightenment and Thomas Jefferson: http://www.monticello.org/streaming/speakers/newton.html.

List of Emblems of Classical Deities in Ancient and Modern Pictorial Arts. Lists identity tags of classical mythological characters in pictorial arts, and also in poetry and fiction. For example, Hercules is often shown with the skin of the Nemean lion, which he killed: http://homepage.mac.com/cparada/GML/003Signed/SREmblems.html.

Littératture française en édition électronique. Books in French from the sixteenth century to the present. http://www.scribd.com/group/449-litt-rature-fran-aise-en-dition-lectronique.

The Newtonian Moment: Science and the Making of Modern Culture. An exhibition organized by The New York Public Library in cooperation with Cambridge University Library: http://www.nypl.org/research/newton.

The Newton Project. Texts and images of many of Isaac Newton's works: http://www.newtonproject.sussex.ac.uk/prism.php?id=1.

The Nine Planets: A Multimedia Tour of the Solar System: http://www.nineplanets.org.

The Online Newton Project. Online access to Isaac Newton manuscripts in the Grace K. Babson Collection, and all of Newton's books that were published in his lifetime in Latin, English, and French. http://www3.babson.edu/Archives/museums_collections/On-Line-Newton-Project.cfm.

Open Source Shakespeare. Complete texts of all plays, sonnets, and poems; character list and search; concordance; word search; text statistics, including word frequencies; and bibliography: http://www.opensourceshakespeare.org.

Out of This World: The Golden Age of the Celestial Atlas. Exhibition of Rare Books from the Collection of the Linda Hall Library: http://www.lhl.lib.mo.us/events_exhib/exhibit/exhibits/stars/welcome.htm.

Quest for Immortality: Treasures of Ancient Egypt. From the National Gallery of Art: http://www.nga.gov/exhibitions/2002/egypt/index.shtm.

Read Print. Thousands of free books, by Shakespeare, Chaucer, Voltaire, H. G. Wells, Jules Verne, Adam Smith, etc.; author index: http://www.readprint.com.

The Solstice Project. Studies, documents, and preserves the Sun Dagger, an ancient Pueblo Indian celestial calendar in Chaco Canyon, New Mexico, and other achievements of ancient Southwestern cultures. Also downloadable research papers and links to other Web sites. http://www.solsticeproject.org.

Star Names: Their Lore and Meaning. Discusses myths related to stars: http://penelope.uchicago.edu/Thayer/E/Gazetteer/Topics/astronomy/_Texts/secondary/ALLSTA/home.html.

The Starry Messenger. Phase 1 of *The Electronic History of Astronomy,* developed in the Whipple Museum of the History of Science and the Department of History and Philosophy of Science, Cambridge University. Subjects include instruments, themes, and personalities; brief essays, images, bibliographies, and overall index. http://www.hps.cam.ac.uk/starry/starrymessenger.

Theoi Greek Mythology: Exploring Greek Mythology in Classical Literature and Art. Profiles of gods and other characters from Greek mythology, with more than one thousand pictures. http://www.theoi.com.

The War of the Worlds Book Cover Collection. Searchable by graphical element, language, date, artist, and publisher; includes interior images. http://drzeus.best.vwh.net/wotw/wotw.html.

Index

About the Authors

EDITH W. HETHERINGTON has a PhD in curriculum and instruction and has taught English and English as a second language at several schools, including Razi University in Iran. More recently, she was admitted to the California Bar and now is the export compliance officer for Chevron Information Technology Company. She has traveled widely in North America, Europe, and the Middle East and has a first-hand acquaintanceship with many of the places mentioned in this book as well as their art and literature.

NORRISS S. HETHERINGTON is the director of the Institute for the History of Astronomy and a visiting scholar with the Office for the History of Science and Technology at the University of California–Berkeley. He is the author of several books, including *Planetary Motions: A Historical Perspective* (Greenwood, 2006). He was a National Endowment for the Humanities Fellow in Studies in Interrelationships between Human Values and Science and Technology.